Springer Climate

Series editor

John Dodson, Menai, Australia

For further volumes:
http://www.springer.com/series/11741

Springer Climate is an interdisciplinary book series dedicated on all climate research. This includes climatology, climate change impacts, climate change management, climate change policy, regional climate, climate monitoring and modeling, palaeoclimatology etc. The series hosts high quality research monographs and edited volumes on Climate, and is crucial reading material for Researchers and students in the field, but also policy makers, and industries dealing with climatic issues. Springer Climate books are all peer-reviewed by specialists (see Editorial Advisory board). If you wish to submit a book project to this series, please contact your Publisher (elodie.tronche@springer.com).

Silvia Macchi · Maurizio Tiepolo
Editors

Climate Change Vulnerability in Southern African Cities

Building Knowledge for Adaptation

 Springer

Editors
Silvia Macchi
Sapienza Università di Roma
Rome
Italy

Maurizio Tiepolo
Politecnico di Torino
Turin
Italy

ISSN 2352-0698
ISBN 978-3-319-00671-0
DOI 10.1007/978-3-319-00672-7
Springer Cham Heidelberg New York Dordrecht London

ISSN 2352-0701 (electronic)
ISBN 978-3-319-00672-7 (eBook)

Library of Congress Control Number: 2014931998

Printed on acid-free paper

Springer is part of Springer Science+Business Media (www.springer.com)

Foreword

I am delighted to introduce this landmark volume focusing specifically on the challenges of climate change in the southeastern African context, where vulnerabilities are high and resources and preparedness still very limited in all but a few specific locations. In a region affected by decades of structural adjustment and deepening poverty, the recent economic upturn, driven in significant part by a resources boom, has renewed foreign investor interest and begun to reverse the legacy of 'de-development'. Its impact, however, remains highly unequal, both geographically and socially, being concentrated overwhelmingly in a few select areas of capital and commercial cities, along narrow transport corridors and (outside of South Africa) in the hands of elites and small professional classes. Despite some promising signs, including of greater governmental transparency and accountability, few of the Millennium Development Goals will be achieved and poverty remains entrenched.

In other words, institutional capacity and public resources generally remain seriously inadequate for addressing the diverse and long-standing development challenges and priorities. Many governments are still seeking to catch up with implementation backlogs from the lean years through flexible combinations of domestic investment and concessionary official development assistance. As climate or broader environmental change began rising rapidly up the global agenda over the last 10–15 years, it is unsurprising that African responses were mostly cautious and even sceptical. Unlike 'climate sceptics' in wealthy countries, however, this had little to do with doubting the science or related projections; rather it reflected a combination of several factors. First, the science—as expressed, for instance, in global climate models—was often not understood. Second, in situations where many people go hungry and meeting basic needs was again being prioritised, there was little obvious political or practical mileage in diverting scarce resources towards addressing problems that were still generally modest but likely to become severe in several decades' time. Finally, it often seemed hypocritical of OECD leaders to be asking (or requiring through aid conditionalities) their counterparts in poor countries to tackle deforestation and greenhouse gas emissions for the global good when OECD countries had grown rich through polluting and resource-intensive industrialisation.

Indeed, this perception has been sufficiently widespread and politically convenient to have underpinned the negotiating position of the G77 group of poor countries over the replacement of the Kyoto Protocol of the UN Framework Convention on Climate Change (UNFCCC). They have been arguing for the establishment of a sizeable global adaptation fund by OECD countries to support climate-friendly and sustainable development in exchange for action to tackle climate change. Nevertheless, over the last decade, most governments have adopted National Adaptation Plans of Action (NAPAs) or similar strategies that demonstrate increasing acceptance of the realities of climate change and the necessity of putting in place frameworks to address it. Recently, too, the East African Community and Southern African Development Community have focused attention on the transboundary nature of many of the key climate and environmental challenges which require regional-scale programmatic interventions (Simon 2012).

One conspicuous feature of virtually all of these national and (sub-)regional strategies is their framing in undifferentiated national terms and then an almost total focus on traditional sectoral and rural interventions, while neglecting or omitting explicit attention to the urban dimension. This is ironic because urban areas are everywhere becoming more important demographically and economically. Even in the less urbanised countries of southeastern Africa, towns and cities represent the densest concentrations of population, resource use, economic opportunity, wealth generation and emissions production. Wealth and poverty stand in sharp, spatially defined juxtaposition and, to be sure, the rich and poor have very different livelihood and emissions profiles. Nevertheless, both are potentially problematic in different ways. While the conventional focus is on car and industrial emissions, in poor urban communities smoke generated by burning wood or charcoal contributes significantly on account of the number of such fires. This smoke is also a major cause of chronic respiratory illness as a result of long-term inhalation in confined spaces within unventilated or poorly ventilated dwellings. The loss of woody biomass in peri-urban and increasingly in distant rural areas to supply urban demand, through what is often now a highly commoditised and large-scale industry, is also contributing to the land-use cover change and the loss of important carbon sinks in those source areas.

Projections of rising temperatures and falling and increasingly variable and unpredictable rainfall across much of Sub-Saharan Africa have become clearer in successive assessment reports of the Intergovernmental Panel on Climate Change (IPCC). Indeed conditions are already changing and are starting to reduce the growth and regeneration rates of woody biomass on which rural and much urban energy demand relies, as well as posing increasing food security challenges, including for urban areas. The complexity of these examples highlights the need for appropriate and collaborative action by a range of institutions and actors, of which urban local authorities are but one—albeit vitally important—category.

Against the background sketched above, a few cities in southern Africa—and especially South Africa's metropoles of Durban, Johannesburg and Cape Town—have taken the lead in assessing, articulating and starting to address the range of issues under the climate change umbrella. They are, therefore, far ahead of the national government. Key to their efforts has been the catalytic role played by well-informed individual 'champions' in senior environmental or political positions able to raise and maintain the focus, and to explain the local implications of broader processes, and how cities would benefit from addressing rather than neglecting climate change through integrating climate change concerns with ongoing 'normal' infrastructural or development activities. Comprehensive accounts of climate change strategies at the city scale are now beginning to emerge through local governments' own efforts in collaboration with university or local consultancy partners (ASSAf 2011; Cartwright et al. 2012).

Such purely endogenous initiatives are not possible for the majority of towns and cities which lack the personnel, institutional capacity and resources for more than incipient efforts (Simon 2010; Carmin et al. 2012). External support through bilateral donors, multilateral agencies such as UN-HABITAT's Cities and Climate Change Initiative (CCCI) and institutional partnerships with foreign researchers is, therefore, essential. Such is the nature of the important and innovative research reported in this stimulating volume, which will surely become a benchmark for understanding the challenges of climate change and how to address them in Dar es Salaam and Maputo, with additional chapters on Dakar and Mozambique's Caia District. As the editors explain in their Preface, the research was undertaken largely under the umbrella of two Italian-funded collaborative projects with local authorities, academics, CCCI staff and others in Dar and Maputo. The chapters cover biophysical dimensions of environmental change, socio-economic assessments of hazard vulnerability to the most important dimensions of climate change (especially flooding and saltwater intrusion), and analyses of coping capacity and the difficulties involved in mainstreaming climate change adaptation. The authors are well versed in the current international literatures from academic and more practice-based sources, and integrate their own research effectively in this context, ensuring that the book does much more than simply local gap-filling in terms of case studies. I am confident that this volume will find its place at the forefront of the expanding literature on assessing and addressing urban climate/environmental change in Africa and beyond.

Egham, UK   David Simon

References

Academy of Science of South Africa (ASSAf) (2011) Towards a low carbon city, focus on Durban. ASSAf, Pretoria

Carmin J, Anguelovski I, Roberts D (2012) Urban climate adaptation in the global south: planning in an emerging policy domain. J Planning Educ Res 32(1):18–32

Cartwright A, Parnell S, Oelofse G, Ward S (eds) (2012) Climate change at the city scale: impacts, mitigation and adaptation in Cape Town. Earthscan, London

Simon D (2010) The challenges of global environmental change for urban Africa. Urban Forum 21(3):235–248
Simon D (2012) Climate change challenges. In Saunders C, Dzinesa GA, Nagar D (eds) Chapter 13, Region building in Southern Africa: progress, problems and prospects. Zed Books for Centre for Conflict Resolution, Cape Town, London, pp 230–248

Preface

This book was written to bridge a gap that we believe is still affecting the scientific literature on climate change in Southern Africa, namely the lack of understanding of hazards and the impact they have on areas prone to them, of the local population's capacity to adapt and local authorities' ability to respond, not to mention the methods used to estimate levels of risk and vulnerability, factors that are useful when planning adaptation to climate change in large cities.

This book is the product of work done by two research teams created by a partnership between Italian universities and African institutions. One of these teams, based at the Interuniversity Department of Regional and Urban Studies and Planning (DIST) of the Politecnico di Torino, is run by Maurizio Tiepolo. Its members include Sarah Braccio, Antonio Cittadino, Magueye Diop, Francesco Fiermonte, Diéthié Ndiaye, Pamoussa Ouedraogo, Enrico Ponte and Stefania Tamea. The other team, based at Sapienza University of Rome's Department of Civil, Building and Environmental Engineering (DICEA), is run by Silvia Macchi and is comprises of 20 researchers from Italy and Tanzania, including Francesco Cioffi, Luca Congedo, Giuseppe Faldi, Laura Fantini, Michele Munafò, Liana Ricci, Matteo Rossi, Giuseppe Sappa, as well as Gabriel Kassenga and Dionis Rugais from Dar es Salaam's Ardhi University.

In May 2010, the two groups joined forces to present the research project entitled *Assessing, Planning and Managing the Territory and the Environment Locally in Sub-Saharan Africa* at the Italian Ministry for University and Scientific Research as part of the call for Research Projects of National Interest (known by the Italian acronym PRIN). The project was co-financed in July 2011 (no. 2009SX8YBH) and developed from August 2011 to October 2013 under the supervision of Maurizio Tiepolo.

In the first few months of work, it quickly became clear that urban vulnerability to climate change was a growing environmental concern in Southern Africa. It was felt that in order to make the comparison clearer, it would be useful to identify case studies involving similar hazards, i.e. sources of potential harm in terms of human injury and damage to health, property, the environment and other things of value. Moreover, when it came to this particular issue, we felt it would be best to choose cases of exposure to effects caused by multiple hazards, such as the floods caused by extreme rains and sea-level rise in dense urban areas, or groundwater salinization resulting from the combined effect of decreasing rainfall and increasing temperature as well as soil sealing in peri-urban areas. The large coastal cities of Maputo, the capital of Mozambique, (Politecnico di Torino) and Dar es Salaam, Tanzania's largest city (Sapienza University of Rome) were therefore selected.

The results of the preliminary studies conducted by the two research teams were presented at the international Urban Impact of Climate Change in Africa (UICCA) conference, organised in partnership with Turin's provincial government on 16 November 2011 in Turin. Other Italian research centres studying climate change adaptation in Sub-Saharan Africa were invited to attend (the IBIMET-CNR National Research Council's Institute of Biometeorology of Italy, Venice's IUAV Istituto Universitario di Architettura, the University of Florence, and the University of Trento), as well as many Italian local authorities and ministerial departments, so as to broaden the discussion increase opportunities for debate and raise awareness of this issue among the many different levels of cooperation (bilateral and decentralized).

After the conference, this book began to take shape and later saw a further opportunity for verification at the international workshop entitled 'Towards Scenarios for Urban Adaptation Planning: Assessing Seawater Intrusion Under Climate and Land Cover Changes in Dar es Salaam', organized at Sapienza University of Rome on 20–22 April 2013.

The contents of this book range from the assessment of risks associated with climate change to the adaptation strategies for reducing vulnerability in two of the most populated cities on the eastern coast of Africa: Dar es Salaam (4.4 million inhabitants) and Maputo (2.4 million). These two main case studies were supplemented by two complementary studies on Dakar (2.9 million) and the Caia district in Mozambique (Fig. 1).

The conceptual frameworks for disaster risk management and climate change adaptation in the scientific literature as well as in those produced by the main multilateral and bilateral development aid agencies are clarified. Next, the assessment methods and applications concerning the various different factors involved are presented, adapted to situations where information is often lacking or where information is scattered and access to it is limited. We believe this is why this book is so ground-breaking compared to the publications currently available on urban adaptation to climate change in Africa. If this achievement has been attained, it is thanks to the in-depth knowledge of sources of information, combined with the great efforts made to fill gaps by obtaining new data, and thanks also to the practical goal of this research, namely to provide urban authorties with the risk analysis and adaptation planning tools necessary to diminish local vulnerability to climate variability and change.

This approach makes the book particularly useful to graduate students, researchers, and practitioners interested in enhancing their knowledge and skills as regards integrating climate change into applied research and development projects in urban Africa.

The book begins with two introductory chapters that review the current state of adaptation to incremental climate stress (Chap. 1) and flood risk reduction and

Fig. 1 The 42 largest cities south of the Sahara in 2010. Over 6 million inhabitants (*very large dot*), 4–5 million (*large dot*), 2–3 million (*medium dot*), 1–2 million (*small dot*) (formulated by E. Ponte based on data published by UN-Habitat's State of the World's Cities 2012/2013. Prosperity of Cities, 2012)

climate change (Chap. 2) in urban studies. The body of the book then presents relevant case studies (Chaps. 3–14), followed by conclusions and recommendations (Chap. 15).

Chapter 1 (Macchi) examines the issue of adaptation planning in cases where incremental stress on systems of natural resources is foreseen due to the combined effects of climate change and a series of other factors of environmental decay, such as urban sprawl. The chapter particularly tackles the vulnerability of access to water caused by the continued degradation of water sources in peri-urban coastal areas of large Sub-Saharan cities. After situating the issue of adaptation within the international discourse on responses to global warming, the specific spatial context under examination is introduced, together with the guiding concept for vulnerability assessment: the adaptive capacity of inhabitants. In addition, three theoretical pillars for adaptation planning are explored: uncertainty as an opportunity for an unfettered vision of the city's future; the centrality of incremental environmental stress in assessment of vulnerability to extreme weather and climate events; and crossing boundaries within science and between science and society for an effective and equitable definition of the problem.

Chapter 2 (Tiepolo) illustrates flood risk reduction following extreme physical events attributable to climate change in large cities south of the Sahara. Large cities are understood here as those with a population greater than one million inhabitants, and the term 'extreme physical events' refers to those events whose likelihood in a given place and time is in the 90th percentile. The main hazard that increasingly hits cities south of the Sahara is flooding. In coastal cities, this is caused by extreme rainfall and sealevel rise. The chapter assesses whether there is enough information available to assert—as the literature currently does—that urban flooding is caused by climate change. The scale of flooding and its impacts are then examined. Finally, the current state of knowledge concerning adaptation measures is presented, with a particular focus on strategies and local adaptation plans. Overall, our results have revealed several commonly held misconceptions in the field of adaptation. In particular, understanding of the mechanisms that cause flooding has proven to be virtually non-existent, a gap that makes it difficult to develop and identify adequate adaptation measures, from early warning to stormwater drainage. In addition, the knowledge of adaptation plans and the development of uniformly applied best practices is less advanced than expected. Adaptation plans have been adopted by few large cities, and those that are in place demonstrate considerable heterogeneity, despite years of support from international organizations promoting best practices.

Once the current state of understanding of the entire Sub-Saharan African region is established, the next two parts examine in detail a few case studies from Southern Africa.

Part II (Chaps. 3–8) presents the research carried out in Dar es Salaam and concludes with a study carried out in Dakar. The six chapters all refer to the conceptual framework for the assessment of vulnerability to climate change developed by the IPCC (see Chap. 1), where adaptive capacity plays a pivotal role. Chapter 3 (Rugai and Kassenga) begins by considering the impact of climate change and the authorities' ability to respond thereto, focusing mainly on highlighting the fact that unchecked and poorly planned expansion of cities is increasing future risk factors as well as the current expenditure on adaptation paid by communities. Chapter 4 (Faldi and Rossi) focuses on seawater intrusion in coastal aquifers. This is a complex phenomenon, due to the combination of natural and human mechanisms, and if aggravated by climate change it could have dramatic consequences, such as impeding the use of the majority of the wells that currently meet human and agricultural consumption needs. Chapter 5 (Congedo and Munafò) investigates urban sprawl, recognising it to be the main non-climatic factor that will accentuate the effects of climate change. This phenomenon affects a great deal of Dar es Salaam's coastal plain, and monitoring it over time is essential if we wish to guide and evaluate the adaptation decisions to be integrated into urban development planning. The key concept of adaptive capacity is the focus of Chap. 6 (Ricci), which uses it to reinterpret certain characteristics typical of the peri-urban area as essential for guaranteeing the spontaneous adaptation of the local population to present and future environmental changes. An analysis of the information collected in the field leads to a framework proposal for supporting local authorities in decision-making on institutional adaptation. Chapter 7 (Macchi and Ricci) discusses the mainstreaming of adaptation into existing plans and programmes related to the urban development and environmental management sectors. This is an approach to adaptation that is approved by international development agencies, but at the same time has limitations that should be tackled in order to apply it in the right way. Finally, Chap. 8 (Biconne) presents a participatory approach to sharing knowledge among urban players on the environmental, social, and cultural dimensions of climate change. This approach has been tested in the peri-urban settlement of Malika, Dakar, demonstrating its potential as a tool in the decision-making processes of urban adaptation planning.

The second half of the book (Chaps. 9–14) discusses the city of Maputo and is supplemented by a study on the Caia district in central Mozambique. The theme of these six chapters is the mapping of flood risk in case of extreme heavy rain and sealevel rise. The final result is a risk digital map derived from a special open source GIS, and an initial adaptation assessment, i.e. an initial identification of the options for adapting to climate change and their evaluation according to a set of criteria. Given the lack of literature on risk assessment methods in cities south of the Sahara, researchers decided to tackle the various components of risk separately so as to leave more space for an in-depth illustration of the methods used. Thus, flood hazards due to extreme rains (Bacci) and sealevel rise caused by climate change (Brandini and Perna) are discussed in Chaps. 9 and 10. They provide an exhaustive examination of the hazard, which is very rarely discussed in publications dealing with flood risk assessment and mapping, as shown in Chap. 2. Chapter 11 (Braccio) presents the various methods that can be used to identify flood-prone areas. Their use depends on local circumstances (the availability of satellite images with a low cloud cover rate immediately following extreme rain, the availability of local surveys of the flooded areas, etc.). The text then goes on to explain the choice made in the case of Maputo, considering the data and the resources available. Chapter 12 (Ponte) illustrates the choice—among the many options available—to use the equation $R = (H * V * E)/A$ to measure risk, and explains how this equation was used in open source GIS to produce the digital, georeferenced, and updatable map of flood-prone areas. The hazard value calculation identified by Bacci, Brandini, and Perna and the adaptation calculation (Tiepolo, see Chap. 13) are quantified and the vulnerability and exposure calculation is described in detail. Chapter 13 (Tiepolo) treats the adaptation baseline (existing adaptation) and adaptation assessment (future) separately. In this case, reference is made to the complexities involved in ascertaining the measures currently in place when working on large cities with vast flood-prone areas (57.4 km^2 in the case of Maputo). Then the text goes on to elucidate the method chosen to identify the future priority measures and examine their distribution over time. This part ends with Chap. 14 (Ianni) and an analysis of the vulnerability of the Caia district (population of approximately 115,000) in central Mozambique. The main focus of this chapter is the vulnerability caused by the local population's loss of access to the land.

Silvia Macchi
Maurizio Tiepolo

Acknowledgments

This book was prepared in two phases, each of which benefited from the contributions of many people.

First and foremost, the international Urban Impact of Climate Change in Africa (UICCA) conference (Turin, 16 November 2011) was made possible thanks to the support of councillor Marco D'Acri, Simonetta Alberico, and Silvana Scarfato (from Turin's provincial government), thanks to the logistical coordination provided by Mario Artuso and the efforts of Andrea Di Vecchia and many other researchers from the IBIMET-CNR National Research Council's Institute of Biometeorology of Italy, as well as Corrado Diamantini (University of Trento) Raffaele Paloscia (University of Florence), Enrico Fontanari and Domenico Patassini (IUAV). We are truly grateful for their cooperation, which laid the foundations for a network of Italian contributors committed to studying and working on urban climate change in Africa.

As for the international workshop held in Rome in April 2013, we would like to thank the Dean of the Faculty of Civil and Industrial Engineering, Fabrizio Vestroni, and the president of The Institute for Environmental Protection and Research (ISPRA), Bernardo De Bernardinis, for their support of the organisation and their promotion of the event; Laura Fantini for logistical coordination and the directors of Sapienza University's Department of Civil, Constructional and Environmental Engineering (DICEA) for their technical assistance; as well as the many colleagues from Italian and Tanzanian universities who offered scientific contributions. Among these, we would particularly like to thank Maria Dolores Fidelibus (Polytechnic of Bari) and Elifuraha G. Mtalo (Bagamoyo University), who provided timely comments and essential feedback on the research reports presented.

The first on-site missions to develop the Dar es Salaam case study were made possible thanks to the financial support provided by Sapienza University of Rome. Much of the initial information was gathered as part of the project entitled Adapting to Climate Change in Coastal Dar es Salaam (ACC DAR), co-financed by the European Commission's DEVCO.

We would particularly like to thank Prof. Gabriel Kassenga and his colleagues at Ardhi University in Dar es Salaam, Tanzania. Without their hospitality and cooperation it would have been impossible to develop the case study.

We are grateful to Dar es Salaam's City Council and the municipalities of Kinondoni, Ilala, and Temeke, as well as the Drill and Dam Construction Agency for the support provided by its directors during various phases of field research.

Last but not least, we are indebted to Loredana Cerbara (IRPPS CNR, the National Research Council's Institute for Research on Population and Social Policies) and Prof. Giuseppe Sappa (Sapienza University of Rome's DICEA) for the guidance provided during the data analysis stage.

The Maputo case study was developed thanks to the cooperation provided in situ by Paulo da Conceição Junior (Município de Maputo) and the organizational support of Ottavio Novelli (Agriconsulting).

Enrico Ponte (Politecnico di Torino) and Milva Pistoni provided editorial support. Ashleigh Rose translated Chap. 1, the entire Part II and the conclusions and was responsible for the English language editing of the entire book.

Finally, we are grateful to Pierpaolo Riva at Springer for his support through the publication process.

                                                                    Silvia Macchi
                                                                    Maurizio Tiepolo

Contents

Contributors

Maurizio Bacci National Research Council, Institute of Biometeorology, Via G. Caproni 8, 50145 Florence, Italy, e-mail: m.bacci@ibimet.cnr.it

Rita Biconne Urban and Regional Planning Department, University of Florence, Via Micheli 2, 50121 Florence, Italy, e-mail: rita.biconne@unifi.it

Sarah Braccio Interuniversity Department of Regional and Urban Studies and Planning, Politecnico di Torino, Viale Mattioli 39, 10125 Turin, Italy, e-mail: sarah.braccio@polito.it

Carlo Brandini National Research Council, Institute of Biometeorology, and LaMMA Consortium for Environmental Modelling and Monitoring Laboratory for Sustainable Development, Via Madonna del Piano 10, 50019 Sesto Fiorentino, FI, Italy, e-mail: brandini@lamma.rete.toscana.it

Luca Congedo Department of Civil, Building, and Environmental Engineering, Sapienza University of Rome, Via Eudossiana 18, 00184 Rome, Italy, e-mail: ing.congedo.luca@gmail.com

Giuseppe Faldi Department of Astronautical, Electrical and Energetic Engineering, Sapienza University of Rome, Via Eudossiana 18, 00184 Rome, Italy, e-mail: giuseppe.faldi@yahoo.com

Elena Ianni Department of Civil and Environmental Engineering, University of Trento, Via Mesiano 77, 38123 Trento, Italy, e-mail: Elena.ianni@ing.unitn.it

Gabriel R. Kassenga School of Environmental Science and Technology, Ardhi University, 35176, Dar es Salaam, Tanzania, e-mail: kassengagr@ gmail.com

Silvia Macchi Department of Civil, Building and Environmental Engineering, Sapienza University of Rome, Via Eudossiana 18, 00184 Rome, Italy, e-mail: silvia.macchi@uniroma1.it

Michele Munafò Italian Institute for Environmental Protection and Research, Via Brancati 48, 00144 Rome, Italy, e-mail: michele.munafo@isprambiente.it

Massimo Perna National Research Council, Institute of Biometeorology, and LaMMA Consortium for Environmental Modelling and Monitoring Laboratory for

Sustainable Development, Via Madonna del Piano 10, 50019 Sesto Fiorentino, FI, Italy, e-mail: perna@lamma.rete.toscana.it

Enrico Ponte Interuniversity Department of Regional and Urban Studies and Planning, Politecnico di Torino, Viale Mattioli 39, 10125 Turin, Italy, e-mail: enrico.ponte@polito.it

Liana Ricci Department of Civil, Building and Environmental Engineering, Sapienza University of Rome, Via Eudossiana 18, 00184 Rome, Italy, e-mail: liana.ricci@uniroma1.it

Matteo Rossi Department of Civil, Building and Environmental Engineering, Sapienza University of Rome, Via Eudossiana 18, 00184 Rome, Italy, e-mail: matteo.rossi@uniroma1.it

Dionis Rugai School of Environmental Science and Technology, Ardhi University, PO Box 35176, Dar es Salaam, Tanzania, e-mail: dionisr@gmail.com

Maurizio Tiepolo Interuniversity Department of Regional and Urban Studies and Planning, Politecnico di Torino, Viale Mattioli 39, 10125 Turin, Italy, e-mail: maurizio.tiepolo@polito.it

Part I
Challenges and Approaches

Chapter 1
Adaptation to Incremental Climate Stress in Urban Regions: Tailoring an Approach to Large Cities in Sub-Saharan Africa

Silvia Macchi

Abstract Research into adaptation strategies to climate change has become a point of great interest for today's urban environmental planners. At the same time, addressing adaptation to climate change in Sub-Saharan cities is an ethical and epistemological challenge. This article presents an approach, developed in the context of a scientific collaboration between an Italian and a Tanzanian university, to adaptation planning in the coastal peri-urban areas of the city of Dar es Salaam. After situating the research within the international discourse on responses to global warming, the specific spatial context of the study is introduced, together with the assumptions that derive from the interpretive key: the adaptive capacity of inhabitants. The three theoretical pillars upon which the approach is based are also explored: uncertainty as an opportunity for an unfettered vision of the city's future; the centrality of incremental environmental stress in assessment of vulnerability to extreme weather and climate events; and crossing boundaries within science and between science and society for an effective and equitable definition of the problem.

Keywords Climate change debate · Urban planning · Peri-urban Africa · Adaptive capacity · Uncertainty

1.1 Planning for Adaptation in Sub-Saharan Cities

Research into adaptation strategies for territories facing the environmental modifications that will be caused by climate change in the near future has become a point of great interest for today's urban environmental planners. Adaptation means anticipating the adverse effects of climate change by adopting appropriate actions to prevent or minimize damage, without forgetting to seize the positive

S. Macchi (✉)
Department of Civil, Building and Environmental Engineering,
Sapienza University of Rome, Via Eudossiana 18, 00184 Rome, Italy
e-mail: silvia.macchi@uniroma1.it

S. Macchi and M. Tiepolo (eds.), *Climate Change Vulnerability in Southern African Cities*, Springer Climate, DOI: 10.1007/978-3-319-00672-7_1, © Springer International Publishing Switzerland 2014

opportunities that may also arise. Among the latter is the opportunity to reconsider mainstream planning models by analyzing their impacts on people's *adaptive capacity*, which in Sub-Saharan cities is closely connected to the direct and informal modalities through which people relate to the natural environment. The ACC DAR[1] research project has pursued this path, and the present article will frequently refer to the project in order to focalize certain open questions posed to urban planning by the demand for adaptation in Sub-Saharan Africa.

Addressing adaptation to climate change in Sub-Saharan cities requires an important effort to combine the prevailing themes in the international discourse on strategies for dealing with global warming and the particular characteristics of the settlement modalities and institutional structures of specific locations. After situating the research with respect to several controversial elements of the relationship between adaptation and mitigation and how to conceive the *what to do* about adaptation, the spatial context of the study is introduced: coastal peri-urban areas of Sub-Saharan cities. Adaptive capacity is the interpretive key that allows for a positive view of a reality that is often described exclusively in terms of what it lacks compared with *real* cities. This view informs the choices at the foundation of this research. In addition, three theoretical elements have contributed to the development of a methodological framework: uncertainty as an opportunity for an unfettered vision of the city's future; the centrality of incremental environmental stress in assessment of vulnerability to extreme weather and climate events; and crossing boundaries within science and between science and society for an effective and equitable definition of the problem. On the basis of this theoretical groundwork, the present research seeks to develop a locally tailored approach to the participatory design of adaption initiatives in the field of urban development and environment management planning.

1.2 Elements of the International Debate Surrounding Adaptation

1.2.1 Adaptation and Mitigation

Article 1 of the 1992 United Nations Framework Convention on Climate Change defines climate change as

---

[1] The Adapting to Climate Change in Coastal Dar (ACC DAR) project is financed by EuropeAid, within the Environment and Sustainable Management of Natural Resources including Energy Thematic Program. It is a three-year project, which will conclude in 2014, that aims to improve the capacity of Dar es Salaam's local governments in local adaptation planning. It is coordinated by the present author and carried out in collaboration with professors and young researchers from Sapienza University in Rome, Italy, and Ardhi University in Dar es Salaam, Tanzania. Chapters 3, 4, 5, 6 and 7 of this book have been developed within the context of that project. All the materials produced in the course of the project are available at www. planning4adaptation.eu.

a change of climate which is attributed directly or indirectly to human activity that alters the composition of the global atmosphere and which is in addition to natural climate variability observed over comparable time periods (UNFCCC 1992).

Although this definition was later revised by the Intergovernmental Panel on Climate Change to include all long-term climate changes irrespective of whether they are due to natural variability or as a result of human activity (IPCC 2001), the international commitment to mitigate climate change through the reduction of its human drivers has become ever stronger over the last two decades. The same cannot be said for adaptation, a strategy that should complement mitigation in meeting the ultimate objective of the UNFCCC. Adaptation refers to

adjustment in natural or human systems in response to actual or expected climatic stimuli or their effects, which moderates harm or exploits beneficial opportunities (IPPC 2001).

In fact, the majority of research and policy efforts have been devoted to stabilizing the concentration of greenhouse gases in the atmosphere (Klein et al. 2005), whereas the commitment to adaptation has become tangible only since the 2007 Bali Conference of Parties under the pressure from the least developed countries who disproportionately suffer the effects of climate change, though they contribute less to GHG emissions than the developed world does.

This slow movement towards an integrated approach that simultaneously addresses mitigation and adaptation concerns has been accompanied by progressively more complex conceptual frameworks for the assessment of vulnerability to climate change (Füssel and Klein 2006) (Fig. 1.1). At the same time, growing attention has been paid to local *adaptive capacity* which is, among other things, the crux of the relationship between disaster risk management and adaptation to climate change.

While the centrality of the notion of adaptive capacity is explored below, a few comments should first be made on the debate surrounding adaptation strategies and their relationship with mitigation. Many believe the efforts of the world's major economic actors in terms of mitigation have clearly been insufficient, given that the temperature of the planet continues to rise, thus scenarios that would render any planned adaptation effort futile have yet to be forestalled. For such individuals, the best form of adaptation is a drastic reduction of greenhouse gas emissions through a radical revision of the dominant development model. To such assertions, the advocates of a greater commitment to adaptation reply that even the most radical change to the development model will not be able to save us from the consequences of the current levels of global warming. In other words, irreversible climate change has already occurred, and adaptation is therefore a necessity. However, the debate is still open as regards the best path forward, whether it is practicable, and how it should be implemented (aside from the widely accepted fact that it should never conflict with mitigation), and it is these questions that the present article shall explore.

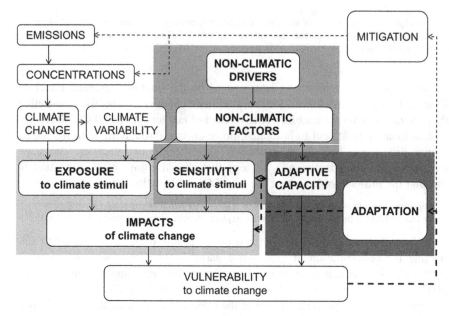

Fig. 1.1 Conceptual framework for the assessment of vulnerability to climate change, with the three areas of interest of the present research highlighted (Modified from Füssel and Klein 2006)

1.2.2 Adaptation: What to Do?

Two positions on adaptation merit particular consideration because they highlight several questions that are crucial for defining the *what to do* of adaptation. Two authors have been selected who have explored those questions by investigating the relationship between climate and human activities, one from a geographical perspective (Parenti 2011) and the other from a historical perspective (Acot 2003).

Parenti's thesis is that assessment of the human impacts of climate change requires an understanding of how changing environmental conditions in the world today are inserted into the *vulnerability trajectory* that is already deeply rooted in the Global South due to the militaristic legacy of the Cold War and the uncontrolled activities of a neoliberal economy. According to Parenti, spontaneous forms of adaptation are already visible, and the common thread among them is violence: in the Global South, ethnic irredentism, religious fanaticism, revolts, xenophobia, banditry, drug trafficking, and local wars for natural resources; in the Global North, counter-insurgency operations, militarization of borders, anti-immigration policies, and xenophobia inspired by the idea of the *armed lifeboat*. In brief, the role of environmental change as regards human activities, whether or not it is caused by climate change, is that of a "drastic accelerant" of the present crisis, "like gasoline on a smoldering fire" (Parenti 2011, p. 65).

Parenti also reminds us of the need to define the *who* of adaptation, to identify the people who will benefit from a given adaptive strategy as well as those who risk becoming less capable of adapting as a result of the effective adaptation of others. By comparing observations of what is occurring in the US with a collection of evidence from the South of the world, the author highlights the conflict between the adaptive capacity of the Global North and that of the Global South. In particular, Parenti demonstrates how the adaptation of the former is currently founded on the capacity to deploy counter-insurgency operations in all four corners of the planet, according to a strategy that aims to destroy the social bonds that are the foundation of adaptive capacity in the latter. As a result, the priority sphere of adaptive action is that of international relations, and the objective is to limit the power of certain actors while simultaneously promoting redistribution of wealth policies in order to render world geography slightly less unequal.

The analysis proposed by Acot integrates Parenti's reasoning into a broader historical reflection on the relationship between environmental change and human activities and, at the same time, leads to slightly different conclusions regarding *what to do*.

Acot agrees with Parenti that the human response to climate changes that occur over a decade has essentially taken the form of conservative adaptation, in line with the trajectory traced by other factors. Only in the case of a longer-term climate change in the order of a century, such as the Little Ice Age that affected Europe during 1550–1850, does the author believe that adaptation can have a transformative impact, though there are always other contributing factors (Acot analyses the French Revolution from this perspective). For both authors, assessment of the effects of climate change cannot exclude consideration of the human factors that are the true drivers of change.

Therefore, though he agrees with Parenti's assertion that choosing the form of adaptation is an essentially political question, Acot warns against defining adaptation strategies on the sole basis of analysis of the present. He argues that, given the relative uncertainty regarding not climate change itself but the ways in which its effects will combine with socio-environmental systems, deciding today what will be needed tomorrow to respond to climate change is closer to a utopian vision than deductive reasoning. In other words, a vision of the future is needed in order to decide which capacities should be protected and which should be developed, and a backcasting approach must be used in order to find solutions to the frightening scenarios that have been outlined using forecasting. The real problem is deciding who should develop this vision of the future, and who should ensure that it comes to pass.

On this last point, Acot notes that political shortsightedness and unequal development of the planet are two heavy weights on any adaptation project. On one hand, there is a substantial difference between ecological timelines and political ones:

what politician would be tempted to make crucial and potentially unpopular decisions when neither he, nor his electorate will see the results? (Acot 2003, p. 260).

On the other hand, certain forms of environmental degradation are "socially irreversible" (ibid.). If, in theoretical terms, any ecosystem in an extreme state of degradation can be rehabilitated, nevertheless

that would require energy inputs that the majority of human society is unable to provide given the current state of indigence of three quarters of the planet (ibid.).

The only way out is a radical cultural change, which Acot defines "the ecology of human liberation" (ibid.).

The reflections of Parenti and Acot have informed our research, and we hope to have made contributions along the lines they have indicated.

1.3 Rethinking Urbanity in Terms of Adaptive Capacity

1.3.1 Peri-Urban Areas as a Starting Point for Redefining Sub-Saharan Urbanity

The joint research project of Sapienza and Ardhi universities began in 2009, and from the beginning the focus has been *peri-urban areas*, understood as a group of settlements and activities where urban and rural characteristics are inextricably interwoven, giving life to a very specific form of territory in terms of lifestyle and relationship types. Such areas are expanding exponentially around one or more densely developed and infrastructured nuclei in many Sub-Saharan cities, to the point that they now occupy entire regions and even exceed their boundaries (see Chap. 5) (Fig. 1.2).

The peri-urban is a growing phenomenon throughout the world—though it occurs under different names—and represents the main modality of urban development in *the century of urbanization*. This is why it is attracting the attention of many urban studies scholars, who are divided into those who see the expansion of peri-urban areas as a threat to the modern and Western project of urbanity defined by the classics of twentieth-century urban studies, from Park to Simmel, and those who, on the other hand, recognize in that expansion an opportunity to review certain paradigmatic assumptions in urban studies and planning practice, which would also respond to post-colonial criticisms (Roy 2009).

The former see the peri-urban as an unfinished city, not suitable to be a "privileged site for the invention, propagation, and cultural experience of modernity" (Robinson 2011, p. 3) because they are rife with shortcomings, like

limited urban infrastructure, informal construction methods, lack of planning, lack of economic opportunity, informal economic activities, large population growth with limited economic growth, [and] external dependency (ibid.).

Fig. 1.2 Peri-urban settlement in the Ilala municipality, Dar es Salaam (photo by Silvia Macchi)

For the latter, most urban growth worldwide in the twenty-first century, while often referred to as an *urban revolution,* should rather be termed as *suburban revolution,* which invites urban scholars and activists to place "the universality of the suburban experience and the boundless divergence in its real processes and outcomes" at the top of their agenda (Keil 2013).

The present research follows the second line of thinking, which it has sought to apply specifically to Sub-Saharan cities through the case study of Dar es Salaam due to the particularly interesting and pressing nature of the challenges of that geographic context. Sub-Saharan cities have been grappling with the definition of their own project for urbanity since the time of decolonization, and have sought to make their way through a variety of obstacles, including the conditions imposed by the colonial legacy in terms of built urban forms and consolidated governmental practices, as well as laws and plans that orient urban development even today. Local planning practice is therefore heavily constrained by the burden of university training that too uncritically continues to propose approaches and instruments that were conceived elsewhere and are saturated with Western urbanity. Lastly, it is difficult for Sub-Saharan cities to resist the pressures from donors and investors when their financial resources are decidedly inadequate to respond to current needs or plan the interventions necessary to accompany population growth rates that exceed 5 % per year (Friedmann 2005). Recognizing that a contextualized definition of what is *urban* is indispensable to Sub-Saharan cities' development of their own specific approach to planning, we expect that analysis of the peri-urban areas of Sub-Saharan cities, specifically Dar es Salaam, will provide the starting points for new ways of thinking about potential types of urbanity.

1.3.2 Adaptive Capacity as the Key to Understanding Sub-Saharan Urbanity

Adaptive capacity is the interpretive key that allows researchers to approach a reality that is distant from their own culture and life experience (Ricci 2011; see also Chap. 6). This concept is fundamental to any attempt to address the theme of adaptation to climate change that does not cede to the temptation to automatically define any urban development pattern that deviates from Western urban ideals as vulnerable.

The adaptation perspective forces emphasis to shift towards understanding human systems and their relationships with the natural environment in order to identify the characteristics that could play a positive role if climate phenomena were to necessitate a transformation of existing structures. Such characteristics represent the components of adaptive capacity, which can be understood as a specific usage of the notion of capacity.

Drawing on the definition of capabilities referred to in Sen's "capabilities approach to development" (Sen 1983), the IPCC defines capacity as

> the combination of all the strengths, attributes, and resources available to an individual, community, society, or organization that can be used to achieve established goals. This includes the conditions and characteristics that permit society at large (institutions, local groups, individuals, etc.) access to and use of social, economic, psychological, cultural, and livelihood-related natural resources, as well as access to the information and the institutions of governance necessary to reduce vulnerability and deal with the consequences of disaster (IPCC 2012, p. 33).

As the IPCC (2012) has clearly stated, the notion of capacity is an integral part of the notion of vulnerability while at the same time exceeding it. In other words, vulnerability must be understood as a relative rather than an absolute lack of capacity. The advantage of this approach is that it allows one "to shift the analytical balance from the negative aspects of vulnerability to the positive actions by people" (ibid.) and it is this shift that is considered promising in the context of the present research.

Given those premises, four research assumptions were defined as follows:

- the livelihoods of peri-urban populations in Dar es Salaam are currently characterized by heavy reliance on locally available natural resources (e.g. access to water is guaranteed by wells that directly access the superficial aquifer) and continuous adaptation of strategies in response to environmental changes (e.g. if a natural resources becomes inaccessible, people search for an alternative source, change activity, or move to another area);

Fig. 1.3 Densification under way in a central area of Dar es Salaam (photo by Silvia Macchi)

- *adaptive capacity* is intrinsically connected to the settlement and relational characteristics that are unique to peri-urban areas (e.g. the possibility of simultaneously conducting agricultural activities, informal commerce, and employed labor in industry or administration renders livelihood strategies extremely flexible) and, in many cases, adaptation entails an inequitable division of labor within both families and the community more generally, as well as exploitation and degradation of the natural environment (maladaptation);
- mainstream strategies for urban development and environmental management imply a substantial modification of the same characteristics that currently guarantee adaptation, i.e. they undermine the current set of capacities, despite the lack of evidence that the proposed settlement models—all of which are oriented towards the densification (Fig. 1.3) and infrastructurization of residential areas according to essentially urban modalities—are capable of promoting and sustaining the development of an alternative set of capacities for future adaptation;
- if the foregoing is true, and considering the degree of effort (not only technical and financial, but also social and political) that would be necessitated by an eventual transformation of peri-urban areas into urban areas, as the mainstream strategy proposes to do, it is crucial that a careful evaluation be conducted of the potential for adaptation associated with current urban development and environmental management practices, and of whether it might not be less onerous

and more effective to intervene to improve such potential by integrating climate change into the existing models (see Chap. 6) rather than promoting the adoption of new models whose efficacy in terms of adaptation has yet to be demonstrated.

1.4 Challenges for Research in Climate Change Adaptation

1.4.1 Decision-Making Under Uncertainty

The approach outlined above is completed by the choice to openly address, rather than circumvent, the uncertainty of predictions regarding the effects of climate change, thus bringing the future projects for the Sub-Saharan peri-urban to the center of the study.

The question of uncertainty has the potential to undermine any proposal of *what to do*, since it renders any type of present evidence and future prediction highly contestable. As the IPCC (2001) has stated, climate change, understood as a problem that has been proven to constitute a serious threat to humanity, places the scientific and political community before a *cascade of uncertainty* that progressively grows from the construction of emissions scenarios, to the definition of the scope of possible human impacts, the response of the carbon cycle, the sensitivity of the global climate, and regional climate change scenarios (Fig. 1.4).

At every stage, the system of reference and the relative predictability characteristics change. They range from highly path-dependent systems, which have a sufficient temporal and structural coherence for the assignment of statistically determined probabilities (bio-physical system), to highly undetermined systems, which are closely connected with contingent conditions for which any probabilistic assertion remains entirely based on subjective judgments (human systems).

This inescapable fact is at the basis of a recent special report by the IPCC (2012), which focuses on the relationship between climate change and extreme weather and climate events, the impacts of such events, and the strategies to manage the associated risks. The report recognizes the necessity of dealing with extreme weather and climate events as decision-making under uncertainty, and offers an important contribution to the definition of an approach to risk management that takes into account the lessons learned in the climate change adaptation field regarding how to address difficult-to-predict futures. Moreover, an important emphasis is placed on the need to adopt a transformative perspective, to face the difficulty of defining shared development projects, and the inevitable ethical and political implications this entails.

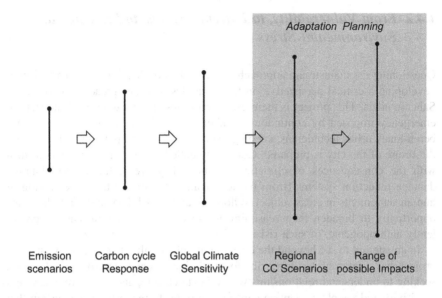

Fig. 1.4 The Cascade of Uncertainty (modified from IPCC 2001)

One example in Dar captures the level of complexity and uncertainty implicated in climate change predictions. In various reports from international organizations on climate change in Dar es Salaam (Dodman et al. 2009; UN-Habitat 2011; START Secretariat 2011; see also Chap. 3) attention is concentrated on coastal areas, for which old fishing settlements, new tourism structures, and the invasion of peri-urban activities and dwellings are currently competing. Thus far everything is *certain*. Rising sea levels are then cited in order to demonstrate that the interventions already on the agenda for protecting the littoral zone from marine erosion are also valid in terms of adaptation to climate change. However, data from the only monitoring station in the Dar region indicate that the sea level has lowered. But that data is only available for the last 20 years, while at least 50 years are needed in order to confirm that this is an irreversible phenomenon traceable to climate change (Kebede and Nicholls 2012). As such, many people attribute uncertainty to the data—20 years are too few—and continue to assume that sea levels will rise in the future. This type of approach that tends to exclude the *uncertain* from the decision-making process is the exact opposite of meeting the challenge posed by uncertainty. The present research project seeks, instead, to focus attention on the *uncertain*. This allows for the recovery of all the knowledge that has heretofore been disregarded, leading to an improved understanding of what the levels of uncertainty really are, and how current urban development and environmental management plans must be modified in order to adequately accommodate them.

1.4.2 From Vulnerability, to Extreme Events, to Incremental Environmental Stress

Questioning the mainstream approach to risk reduction opens up the possibility of developing a critical perspective on the dynamics that are spreading across Sub-Saharan cities. This project is therefore a conscious attempt to cast off the state of emergency imposed by continuous weather-related disasters, which has encumbered many urban technicians, keeping them away from the planning tables where the future of the city is prepared. Constantly called upon by governments to deal with the consequences of climatic extremes, they are restrained to planning damage reduction systems (from barriers against sea storms to the relocation of human settlements in areas subject to flooding and landslides), and rarely have the opportunity to broaden their reasoning to consider the structural causes, prevalently anthropogenic, of such risks.

In an attempt to see beyond the question of vulnerability to extreme events, this project has preferred to direct its focus to incremental environmental stress, seeking to understand relationships with climate change, and therefore ecosystem sensitivity and people's adaptive capacity, in order to reconstruct the system that connects environmental degradation with vulnerability to extreme events. In particular, project researchers have investigated the salinization of the coastal aquifer (see Chap. 4), a phenomenon that is already under way and is a major concern for households living in peri-urban neighborhoods within the coastal plain. In fact, most people depend heavily on boreholes for access to water for domestic and productive (mostly agriculture-related) purposes. Notwithstanding the social importance of groundwater availability, studies on seawater intrusion in the coastal shallow aquifer are scarce and none have developed future scenarios under climatic and non-climatic changes, to our knowledge. Consequently, several research studies have been undertaken to assess changes in past recharge and pumping rates, develop a hydrogeological balance, downscale global climate change predictions to the regional level, and explore the complex interplay between urban sprawl and climate change adaptation, since land cover change might be the most important non-climatic factor that influences future groundwater availability (ACC DAR 2013).

The relationship between incremental stress and vulnerability to extreme events has been well documented by the IPCC (2012):

> The environmental dimension of vulnerability also deals with the role of regulating ecosystem services and ecosystem functions, which directly impact human well-being, particularly for those social groups that heavily depend on these services and functions due to their livelihood profiles.
> [...]
> The degradation of ecosystem services and functions can contribute to an exacerbation of both the natural hazard context and people's vulnerability. The erosion of ecosystem services and functions can contribute to a decrease in coping and adaptive capacities in terms of reduced alternatives for livelihoods and income-generating activities due to the degradation of natural resources. Additionally, a deterioration of environmental services

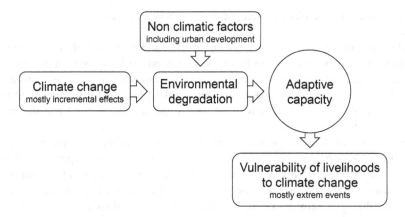

Fig. 1.5 The relationship between incremental stress and vulnerability to extreme events

and functions might also increase the costs of accessing those services, for example in terms of the increased time and travel needed to access drinking water in rural communities affected by droughts or salinization.

Furthermore, environmental vulnerability can also mean that, in the case of a hazardous event, the community may lose access to the only available water resource or face a major reduction in soil productivity, which then also increases the risk of crop failure (IPCC 2012, p. 77).

In other words, a natural system under stress is more sensitive to climate change, while at the same time providing fewer environmental services to inhabitants, thus reducing their adaptive capacity and rendering them more vulnerable to extreme events (Fig. 1.5). If this is true, incremental environmental stresses produced by the combined effect of climate change and non-climatic factors, such as urban development, may prove to be less visible but certainly more insidious than extreme events.

1.4.3 Crossing Disciplinary Boundaries

The third challenge that we have sought to meet is perhaps the most difficult, and continues to raise questions. All the literature on adaptation and sustainable development favors the crossing of disciplinary boundaries, and those who specialize in urban planning are already accustomed and in fact trained to do so. Nevertheless, every case study and every context necessitate research into specific cross-disciplinary methods. How can the questions that emerge from the case study be linked with the competencies of the working group? How does one decide when it becomes necessary to require contributions of disciplines that are not represented in the working group? Presented below are our attempts to go beyond the limits of the field of environmental engineering, already vast and multifarious in its own right, in order to establish a dialogue with human sciences and art, particularly theatre.[2]

The main reason in favor of an inter-, multi-, and trans-disciplinary approach lies in the very nature of the problem to be addressed. As we have seen, adaptation to climate change involves a plurality of interconnected human and physical factors that act on different spatial and temporal scales. To fully understand the complexity of such interaction is a task that surpasses the sectoral and thematic divisions typical of academia.

Moreover, the knowledge needed to formulate the problem is not only of a scientific or *expert* type, but also consistently involves the life experience of the people and value systems of a society. As such, the crossing of disciplinary boundaries is not sufficient; we must go further still to find the instruments that will allow for dialogue between science, policy, technique, and society (see Chap. 8).

This question is particularly evident with respect to the evaluation of people's adaptive capacity and the identification of contextual factors that render them more or less vulnerable. To understand personal-societal-environmental relationships in the specific context in question, one must solve the puzzle generated by the juxtaposition of a plurality of images produced in a variety of ways (in our case: statistical analysis of household questionnaire data, open interviews with institutions and inhabitants, participatory theatre workshops at the community level, land cover analysis through satellite images, hydrogeological analysis and monitoring of the superficial aquifer, etc.).

Learning to master such an analytic approach has proven to be a truly difficult challenge, as it requires each researcher to question his or her methods and open a space for other types of knowledge. It has therefore been an important effort that, moreover, is little rewarded in academic and professional spheres, which explains people's resistance to spending resources on efforts considered to be unprofitable and at high risk of failure.

1.5 Future Prospects

The theoretical elements presented in this article are the product of an on-going research project initiated in 2009 and their application to the case of Dar es Salaam is still under way. This book presents the results obtained thus far, but only after the 2014 conclusion of the ACC DAR project will it be possible to evaluate the efficacy of the proposed approach and reflect on its practicability.

The lessons learned to date (Macchi et al. 2013) provide the basis for developing a locally tailored approach to the participatory design of adaption initiatives

---

[2] The results of this hybrid collaboration are not presented in the present volume because they require further work, particularly reflection on experiments already conducted, but also further experiments. Nevertheless, some materials that present what has been accomplished thus far are available at www.planning4adaptation.eu.

in the field of urban development and environment management planning. In particular, those lessons are being used to improve the use of scenario analysis when dealing with uncertainty in climate change related planning (Faldi 2013). The risk intrinsic to applying the forecasting approach to adaptation is that it favors a conservative definition of planning objectives, as though the purpose of adaptation were to conserve the status quo, which leads to adjustment of the current development model to avoid certain unwanted and bothersome changes (Parenti 2011). The backcasting approach would be an antidote to such risk, which is crucial for cities whose rapid pace of development could provide unique opportunities to shape an optimal future. The next step in the research project will be to explore, in practice, the implications and advantages of using a participatory backcasting approach for planning adaptation.

In addition, we are happy to note that we are not the only researchers following this path, and in fact many others are working on the same premises (see Chap. 8). This growing momentum has great potential to provide an innovative impulse to urban environmental planning, certainly in Sub-Saharan Africa and probably also in the rest of the world, including the Global North.

References

ACC DAR Project (2013) In: 2nd International workshop: towards scenarios for urban adaptation planning. Assessing seawater intrusion under climate and land cover changes in Dar es Salaam, Tanzania. http://www.planning4adaptation.eu/. Accessed 28 Jun 2013

Acot P (2003) Histoire du climat. Perrin, Paris

Dodman D, Kibona E, Kiluma L (2009) Tomorrow is too late: responding to social and climate vulnerability in Dar es Salaam, Tanzania. Case study prepared for UN-Habitat, cities and climate change: global report on human settlements 2011. Available via http://www.unhabitat.org/grhs/2011. Accessed 28 Jun 2013

Faldi G (2013) The use of backcasting scenario for planning adaptation to climate change in sub-Saharan urban areas. Paper presented at the AESOP-ACSP joint congress, Dublin, 15–19 July 2013. Available via http://www.planning4adaptation.eu/. Accessed 15 Jan 2014

Friedmann J (2005) Globalization and the emerging culture of planning. Prog Plann 64(3):183–234

Füssel HM, Klein RJT (2006) Climate change vulnerability assessments: an evolution of conceptual thinking. Clim Change 75(3):301–329

IPCC (2001) Impacts, adaptation, and vulnerability: contribution of working group II to the third assessment report of the IPCC. Cambridge University Press, Cambridge

IPCC (2012) Managing the risks of extreme events and disasters to advance climate change adaptation. A special report of working groups I and II of the intergovernmental panel on climate change. Cambridge University Press, Cambridge

Kebede AS, Nicholls RJ (2012) Exposure and vulnerability to climate extremes: population and assets exposure to coastal flooding in Dar es Salaam, Tanzania. Reg Environ Change 12(1):81–94

Keil R (2013) A Suburban Revolution? An international conference on bringing the fringe to the centre of global urban research and practice. Call for papers. The City Institute at York University (CITY), Toronto http://suburbs.apps01.yorku.ca/2013-mcri-conference-a-suburban-revolution/. Accessed 28 Jun 2013

Klein RJT, Schipper ELF, Dessai S (2005) Integrating mitigation and adaptation into climate and development policy: three research questions. Environ Sci Policy 8(6):579–588

Macchi S, Ricci L, Congedo L et al (2013) Adapting to climate change in coastal Dar es Salaam. Paper presented at the AESOP-ACSP Joint Congress, Dublin, 15–19 July 2013. Available via http://www.planning4adaptation.eu/. Accessed 15 Jan 2014

Parenti C (2011) Tropic of Chaos: climate change and the new geography of violence. Nation Books/Perseus, New York

Ricci L (2011) Reinterpretare la città sub-sahariana attraverso il concetto di *capacità di adattamento* (Reinterpreting Sub-Saharan cities through the concept of *adaptive capacity*). PhD thesis. Sapienza University, Rome http://padis.uniroma1.it/handle/10805/1375. Accessed 28 Jun 2013

Robinson J (2011) Cities in a world of cities: the comparative gesture. Int J Urban Reg Res 35(1):1–23

Roy A (2009) The 21st-century metropolis: new geographies of theory. Reg Stud 43(6):819–830

Sen AK (1983) Poverty and famines: an essay on entitlement and deprivation. Oxford University Press, Oxford

START Secretariat (2011) Urban Poverty & climate change in Dar es Salaam, Tanzania: a case study. Final report prepared/contributed by the Pan-African START Secretariat, International START Secretariat, Meteorological Agency and Ardhi University, Dar es Salaam http://start.org/download/2011/dar-case-study.pdf. Accessed 12 Jun 2013

UNFCCC (1992) Text of the United Nations framework convention on climate change. http://unfccc.int/files/essential_background/

UN-Habitat (2011) Cities and climate change: global report on human settlements 2011. Earthscan, London

Chapter 2
Flood Risk Reduction and Climate Change in Large Cities South of the Sahara

Maurizio Tiepolo

Abstract In the region south of the Sahara, flooding is the most common natural hazard. Large cities are increasingly affected: 7 in the 1980s, 27 in the 1990s, 37 in the 2000s, and 7 more since 2010. Although many studies link this trend to climate change, our understanding of natural events is still too fragmentary to allow us to appreciate the correlation between precipitation and floods. When attempting to bridge this gap, we should also consider how rainfall affects a city's entire watershed (which can be quite extensive). By contrast, the impact of flooding on places, goods, and people is much better understood. Such effects can be so extreme as to bring the economies of large cities to their knees. Over time, the concept of risk as a product of hazard and vulnerability has expanded to include exposure and climate change adaptation. Mapping flood risk is the first step towards identifying adaptation measures, yet only one of the 11 large cities most affected by floods has a detailed flood risk map. Three of the 11 have adopted climate change adaptation strategies and plans. The remaining eight cities use an array of tools whose impact on flood risk reduction is not yet detectable.

Keywords Flood risk · Risk mapping · Climate change · Local adaptation plans · Sub-Saharan cities

M. Tiepolo (✉)
Interuniversity Department of Regional and Urban Studies and Planning,
Politecnico di Torino, Viale Mattioli 39 10125 Turin, Italy
e-mail: maurizio.tiepolo@polito.it

S. Macchi and M. Tiepolo (eds.), *Climate Change Vulnerability*
in Southern African Cities, Springer Climate, DOI: 10.1007/978-3-319-00672-7_2,
© Springer International Publishing Switzerland 2014

2.1 Introduction

Floods account for half of the disasters recorded since 1981 in Sub-Saharan Africa, and 78 of them have hit large cities (EM-DAT). This figure alone renders consideration of flood risk a necessity in the context of adaptation to climate change. In addition, flood risk is becoming an increasingly pressing issue for two reasons: firstly, the number of floods is increasing, presumably due to climate change; secondly, large cities south of the Sahara are growing in number. The extent of potential damage is therefore increasing.

This chapter aims to outline the current understanding of the relationship between floods and climate change in an urban context, and to identify the risk mapping methods and main adaptation measures that are slowly being adopted.

In the past, efforts to reduce flood risk have involved at least two different approaches: disaster risk reduction (DRR), which identified exposed areas, mitigation, and post-event measures; and climate change adaptation (CCA), which identified climate change trends (CC) and measures designed to mitigate the impacts of future flooding.

The possibility of combining these approaches has often been suggested (Mitchell and Van Aalst 2008) and, more recently, largely hybridized studies have drawn on both approaches.

The second section reviews what is known about climate change (precipitation and sea level rise) in urban environments. We then move on to consider the present understanding of the flooding that affects large cities in Sub-Saharan Africa in order to highlight existing knowledge gaps. The third section addresses the impact of flooding on large cities in order to determine whether specific adaptation measures are needed in such areas. The fourth section addresses flood risk reduction, with a focus on maps of risk areas, an aspect that is specifically considered in the case studies collected in this publication, as well as adaptation planning.

In order to complete this process, we drew predominantly from the EM-DAT International Disaster Database and maps of flooded areas produced by UNITAR and the ICSMD. The second and third sections proceed on a case-by-case basis, starting with the large cities that are more exposed to flooding, and draw mainly from reports and plans. When considering the actions taken by international donors to promote risk reduction and climate change adaptation, the main source of information is the literature produced by multilateral organizations. We also reconstruct how the definition of risk has evolved over the past two decades.

2.2 Extreme Rainfall and Flood Risk in Large Cities South of the Sahara

Floods are usually caused by the overflowing of large rivers, by flash floods from their tributaries, runoff following intense local rain, and sea level rise, as well as ground water floods and artificial systems failures (Bloch et al. 2012, p. 27).

In the next three decades, temperatures in Southern Africa are expected to rise and rainfall in Eastern Africa is expected to increase (including in the Horn of Africa). Global mean sea level is expected to rise from 18 to 59 cm (according to different temperature change scenarios) over the next 100 years (IPCC 2007, p. 45, 2010).

Detailed studies of climate change and flooding in vast areas are possible, and those that have been carried out have detected—as is the case in South Africa—changes in the intensity of extreme rainfall events (Mason et al. 1999). However, the current understanding of the link between climate change and floods is lacking with respect to individual large cities (Brinkmann and von Teichman 2010). If it were possible to detect variations in the frequency and intensity (mm/hour) of extreme rainfall in large cities, and changes in the duration of the rainy season, we would be able to assert that we are witnessing climate change. Unfortunately, the large cities where rainfall analyses have been conducted are still too few. Moreover, existing analyses are usually limited to annual rainfall over time (Table 2.5). The large cities where the intensity and frequency of heavy rains has been investigated (Addis Ababa, Dar es Salaam, Douala, Maputo) are exceptions to the rule (De Paola et al. 2013) (see Chap. 9). As a result, climate change predictions are formulated only at the national scale, and then presumed to be applicable to individual cities (UN-Habitat 2010a, p. 7). In its *Global Report on Human Settlements 2011—Cities and Climate Change*, UN-Habitat introduces the efforts made by cities to tackle climate change and its impacts, but it does not discuss the changes in precipitation in the watersheds to which they belong or how this contributes to flooding. This information is essential when planning adaptation measures such as drainage canals.

Moreover, urban climate change studies are limited by the lack of weather stations recording atmospheric conditions (Table 2.1). Thus, the link between extreme weather and flooding in urban centers is little understood, and our knowledge of urban flooding is based only on its impact.

2.2.1 Floods in Large Cities

The Emergency Events Database of the CRED EM-DAT Centre for Research on the Epidemiology of Disasters has been recording disasters worldwide in partnership with the WHO and other organizations since 1988, and makes that record available on the Internet. Over the past 33 years, 654 floods have affected 38 million inhabitants and have claimed almost 13,000 lives in Sub-Saharan Africa. This is the number one hazard in terms of frequency and impact in the subcontinent (almost 50 % of disasters are floods).

The same source reveals that 11 % of disasters hit large cities and cause 12 % of total damage. This figure is approximate, as some events are not reported while others involve not only urban areas, but also wider territories that can include smaller cities or towns. It is worth noting that of the 78 floods that have hit large cities, two-thirds have recurred in the same place, including Bangui (7 times),

Table 2.1 Large cities south of the Sahara, 2013: weather station density

City, year	Area		Weather station
	Built-up area km^2	Administrative area km^2	No.
Dar es Salaam, 2011	400	1,690	1
Abidjan, 2009	–	487	1
Accra, 2000	344	3,245	1
Maputo, 2011	182	347	1
Dakar, 2009	156	580	1
Douala, 2013	–	210	1
Niamey, 2013	122	528	2

(Saley et al. 2009; Angel et al. 2005; République du Sénégal 2010) (see Chaps. 5, 11 and 12)

Djbouti (5 times), Accra and Lagos (4 times each), and Bamako, Cape Town, Dakar, Kinshasa, and Luanda (3 times each). The effect of repeated flooding is that it "will reduce the capacity of communities and others to prepare for these events, respond to them, and rebuild in their wake" (UNEP/OCHA 2012, p. 7).

2.2.2 Trends in Flooding

The number of cities hit by flooding is increasing: from 7 in the 1980s, to 27 in the 1990s, to 37 in the 2000s, and in the first two and a half years of the present decade there have already been 7. The number of people affected rose from 310,000 in the 1980s to 3.8 million in the following decade, then fell to 1.2 million in the 2000s, and has dropped to 0.1 million in the past two and a half years.

Maps of the areas flooded following extreme natural events are increasingly available. UNITAR-UNOSAT produces small-scale maps of flooded urban areas. Maps of disasters as they unfold are also produced by the European Union's Joint Research Centre Global Disaster Alert and Coordination System, as well as by the International Charter on Space and Major Disasters (ICSMD), together with Strasbourg's SERT.

This documentation does not provide information on the size of flooded areas, though it would be easy to determine using the GIS used to generates such maps, and that information would be extremely useful for planners when estimating damage and preparing for recovery work.

The information gathered on the flooded areas of some particularly exposed cities (Table 2.2) shows that the extent of the areas hit varies enormously from case to case: sometimes it is less than 9 % of the built-up area, while in others it is over 50 %.

In large cities, the flooding of small areas affects tens if not hundreds of thousands of people. The number of homeless people can become a veritable nightmare for local authorities, particularly when one considers that floods are not

Table 2.2 Large cities south of the Sahara: a survey of areas exposed to flooding

	Bangui	Banjul	Dakar	Djibouti	Maputo	Niamey	Ouagadougou	Saint Louis
Year of flood	2009	2009	2009	2001	2011	2012	2009	1997
Built-up (ha)	9,250	4,948	15,560	–	18,240	11,400	21,750	386
Flooded (ha)	800	–	800	1,200	–	2,746	14,878	13
Flood risk (ha)	–	60	–	1,333	5,734	–	–	–
P total (thousands)	1,000	35	...	...	1,100	1,100	308[a]	...
P flooded (thousands)	181		360	–	70	84	24[a]	–
Flooded/Built up (%)	9	–	5	48	–	24	68	3
Flood risk/Built up (%)	–	1	–	53	31	–	–	–
P flooded/P total (%)	18		8	–	7	8	8	–

(Nguimalet 2007; Jaiteh and Sarr 2010; Angel et al. 2005; République de Djibouti et al. 2001; Braccio and Tiepolo 2013; Ville de Niamey 2012; République du Burkina Faso 2010; UNITAR 2009) (see Chap. 12)
[a] Households

brief episodes: water can often linger for months, and where houses are built of earth (as is still the case in many cities), they collapse.

Over half of the large cities hit by flooding are located along the coast (Fig. 2.1). This means that they are also exposed to the risk of flooding due to a rise in sea level.

There is no connection between the 2013 human development index and the number of people affected by flooding in urban areas, i.e. there is no proof that countries with lower human development suffer more from flooding. This means that even the most developed countries south of the Sahara do not have sufficient adaptation measures in place to reduce the risk of flooding.

2.2.3 Reasons for Flooding

We don't know what natural events cause each of the disasters recorded by the EM-DAT. In some cases, flooding occurs as flash floods, which are intense and last a short time. In many riverside cities, flooding occurs when the river rises above the level of its banks and not because of extreme rains falling on the city, although rains can contribute to the phenomenon. This type of disaster is generated upstream, by intense rains falling on areas that are sometimes hundreds of miles away from the city, affecting vast watersheds and resulting in the swelling of the river far above its usual level. In the Sahel area, flooding occurs due to erosion

Fig. 2.1 Large cities south of the Sahara, 1981–2013: cities hit by flooding five times or more (*large bullet*), 3–4 times (*medium bullet*), 1–2 times (*dot*). *Source* CRED EM-DAT, formulated by E. Ponte

caused by the reduction of vegetation in higher parts of the watershed (in favor of agriculture and pasture or due to deforestation), which has altered the structure of riverbanks and riverbeds with sand deposits (Mahé 2006). Such was the case with the flooding of Niamey in 2012. Examination of climate change effects on flooding requires an expanded perspective, focusing not only on urban weather stations but considering the network of weather stations throughout the entire river basin upstream of the city, in areas that are, at times, truly vast. The field literature currently available is not an adequate basis upon which to understand such mechanisms.

2.3 The Impact of Urban Flooding South of the Sahara

Flooding has a direct impact on the population, buildings, livestock, crops, and goods, as well as an indirect impact in human, economic, social, financial, political, and institutional terms.

The impact on homes can be devastating if floodwater stagnates. In several parts of Lagos, water from the most recent floods stagnated for a length of time that varied from four to six months (Aderogba et al. 2012, p. 420). Where buildings are made of earth, homes literally melt away. Where they are made of more resistant materials, they remain inaccessible for several days, sometimes weeks. In both cases, properties are devalued (Commissioner for environment 2012). If flooding invades latrines, as in Lusaka (Heat et al. 2010) and in many other cases, the sludge seeps out, contaminating the stagnant water or open fresh water drinking wells, and this can create serious health problems. Indirect problems in this case include the accommodation of displaced persons in camps and their subsequent resettlement.

Another severe impact concerns transport infrastructure such as roads and railways (Adelekan 2000). Their paralysis stops all economic trade, and if petrol distribution is hindered as a result, interruptions in the electricity supply may also occur (UN-Habitat 2011, p. 72).

At other times, flooding can disrupt the provision of drinking water via the municipal supply network (UN-Habitat 2011).

The impact of flooding can be particularly severe in peri-urban agricultural areas (Kolawole et al. 2011), especially in the Sahel, where the great drought of the early 1970s led to the creation of vast irrigation perimeters in order to guarantee food security in urban areas should there be other droughts in future. If high waters exceed the riverbanks and flood the paddy fields, as occurred in Niamey in August 2012, the harvest will be damaged and the country will have to resort to international food aid.

In more general terms, flooding increases the incidence of malaria, can expose the population to toxic substances, harms health, and increases requests for medical assistance.

The impact of flooding on people's livelihoods and goods is potentially huge if settlements are not protected from high water, if they have been constructed in dry riverbeds or if the population has not been warned of the risk of high water with enough warning.

The proliferation of informal settlements with an insufficient number of roads hinders access to relief during emergencies (UNEP/OCHA 2012), increases the impervious surface and therefore runoff, and hinders drainage maintenance (blocked by rubbish). Consequently, an average amount of rainfall may be enough to cause a flood.

The efforts made to react to flooding are greater the more densely populated the exposed area and the more severe the consequences, such as contamination of the clean water supply and treatment centers or the circulation of waste and pollutants

(UNEP/OCHA 2012, p. 17). In any case, it is impossible to evacuate the population of an entire district, especially if there is no contingency plan.

2.4 Flood Risk Reduction in Large Cities

A disaster risk management approach involves the planning, creation, and evaluation of strategies and policies, as well as the organization of measures designed to prepare for, and respond to, emergencies, and to rebuild (IPCC 2012, p. 34).

2.4.1 Recognizing the Risk

The first step is always to recognize the risk. Today, risk of disaster is

> the likelihood of severe alterations in the normal functioning of a community due to hazardous physical events interacting with vulnerable social conditions, leading to widespread adverse effects that require immediate emergency response to satisfy critical human needs... (IPCC 2012, p. 32).

The creation of risk maps is considered a priority, and they should be updated and verified in partnership with potentially exposed communities (UNEP/OCHA 2012, p. 11). From the mid-1990s on, risk was mainly calculated as a product of hazard and vulnerability, and it was listed as such in the manuals of the main multilateral organizations until the mid-2000s. At the same time, towards the end of the 1990s, an approach to risk that initially considered exposure, hazards, and vulnerability, and, more recently, capacity, adaptation or preparedness, made headway, and the manual definition of risk evolved accordingly (Table 2.3).

Although there is general agreement on the determinants of risk, each factor may be ascertained using different sets of indicators, depending on the chosen methodology.

The first step in assessing flood risk is identification of flood-prone areas (Table 2.4). This can be achieved using two different methods. The first involves the observation, recording, and mapping of flooded areas. This method is used by local authorities, particularly after flooding, in order to determine the number of people affected for the purpose of distributing compensation. Many such investigations that apply this method have also drawn information from the personal recollections of local inhabitants. Insofar as past flooding is linked to the amount of precipitation in the watershed, these approaches can also be used to identify flood risk areas for preventative purposes. The second type of method is preventative and is based on hydrological models of the areas that could be flooded by heavy rains. This method is not frequently used in Sub-Saharan Africa due to the amount of time needed to gather information, and the cost. Furthermore, it seems to produce rather imprecise results for urban areas characterized by informal

Table 2.3 Different ways of conceiving of risk and some of its applications

Formula[a]	Origins	Multi/bilateral organisation manuals
$R = H * V$	Blakie et al. (1994)	UNDHA (1992)
	Wisner et al. (2003)	GTZ (2002)
		UN-Habitat (2004)
		IATF CCDRR (2005)
		UNDP 2004
$R = Pr * Co$	Jones and Boer 2003	
$R = H, V, E$	Crichton (1999)	ADRC 2005
	Granger et al. (1999)	
	Turner 2003	
	Dilley et al. (2005)	
$R = H * V * DP$	Villagrán de León (2004)	
$R = H * V * Co$	Kaynia et al. (2008)	
$R = (H * V * Va)/P$	De La Cruz-Reyna 1996[a]	
$R = H + V + E -$ Cc	Davidson (1997)	IADB et al. 2003
$R = H (E + S - A)$		UN-Habitat (2010a)
$R = (H * V)/C$		UN/ISDR 2002
		World Bank (2009)
		USAID (2010)
$R = (H * V * E)/A$	Gotangco and Perez (2010), see Chap. 12	

[a] C = Capacity (= adaptation), Cc = Coping capacities, Co = Consequence, DP = Deficiencies in preparedness, P = Preparedness, Pr = Probability, Va = Value threatened. For terms and definitions see UNISDR 2009

Table 2.4 Large cities south of the Sahara: methods used to identify flood-prone areas

Method	Accra	Gwagwalada	Kano	Maputo
Wet area after heavy rains				●
Elevation of flooded area	●	●		
Census enumeration areas nearest stream channels	●			
Stream buffer zone according to slope angles			●	

(Okyere et al. 2012; Rain et al. 2011; Ishaya et al. 2009; Orok 2011) (see Chap. 12)

settlements, which do not lend themselves well to modeling due to the constant changes they undergo (construction work, blockage of the drainage canal network, increasing soil impermeability, etc.).

A hazard as the threat of or potential for adverse effects comes about when a physical event (rain) reaches a human settlement, and this should be expressed as a rain recurrence interval (Table 2.5). However, this is rarely done.

Those exposed to the hazard will be adversely affected by it according to their level of vulnerability, i.e. their ability or inability to cope with the adverse effects

Table 2.5 Large cities south of the Sahara: hazard appreciation

Physical event characteristics	Abidjan	Addis Ababa	Dakar	Dar es Salaam	Kano	Lagos	Maputo	Niamey	Ouagadougou
Yearly rainfall	•	•	•		•	•			•
Monthly rainfall		•	•		•				
Daily rainfall				•			•	•	
Years considered in rainfall, n°	29	51	40	53	25	105	46	10	30
Recurrence interval of extreme rains							•		
Seasonal variation of tides							•		
Sea level rise due to climate change							•		

(Saley et al. 2009; Conway et al. 2004; Da Costa and Kandja 2002; De Paola et al. 2013; Nabegu 2009; World Bank 2009) (see Chaps. 9, 10 and 11)

Table 2.6 Large cities south of the Sahara: indicators used to assess vulnerability to flooding

Sector	Indicator	Accra	Addis Ababa	Maputo	Lagos
Physical	Soil			•	•
	Impervious surface			•	
	Road			•	
	Tree cover	•	•	•	
	Dumping site			•	
	Elevation	•			
	Lack of drainage			•	
	Housing density			•	•
Socio-economic	Poverty			•	•
	Illiteracy	•			
	Casual laborers	•			
Demographic	Fertility	•			•
	Young	•			•
	Migrants	•			
Health	Mortality under 5 years	•			

(Jankowska et al. 2011; Tamru 2002; Adedeji et al. 2012) (see Chap. 12)

of climate change (IPCC 2012). Vulnerability, in turn, can be broken down into sensitivity, adaptive capacity, and resilience (Table 2.6).

Adaptation—understood not as a collection of existing measures, but rather as a combination of future measures—involves a rainwater drainage system in all large cities considered (Table 2.7). Given that drainage systems often do not have the capacity to meet requirements, and are also obstructed by rubbish left in the street and transported downhill by runoff, rubbish removal is an adaptation measure that is often recommended in urban flooding situations. Also common is the use of flood risk assessment, community awareness, and resettlement of the local population residing in flood-prone areas.

2.4.2 Planning Flood Risk Reduction and Adaptation to Climate Change

The abovementioned measures should be coordinated and applied using a timetable. In other words, they should result from a plan. Local authorities should mobilize awareness of such an instrument even at levels above the local sphere.

UN-Habitat is one of the multilateral organizations committed to climate change adaptation in Africa. Since the 2007 *Global Report on Human Settlements*, UN-Habitat has been confronting the issue of disasters and climate change. Rather than being described, climate change is presumed in its various guises (tropical cyclones, heat waves, floods, and sea level rise). Among its noted consequences is

Table 2.7 Large cities south of the Sahara: recommended adaptation measures for flooding

Adaptation measures	Accra	Addis	Bangui	Durban	Lagos	Maputo	Niamey
Flood risk assessment	•			•	•		
Land use zoning					•		
Coastal set back lines				•			
Contingency plan for risk areas				•			
Master drainage plan				•			
Early warning system				•			
Rainfall/runoff projections				•			
Research into CC impact on floods				•			
Disaster relief fund and centers					•		
Protective margins for infrastructure					•		
Drainage	•	•	•	•		•	•
Drainage maintenance				•		•	
Pumping station to provide drainage					•		
Fight against silting-up of river(s)							•
Soil conservation							•
Slope stabilization/riparian vegetation				•		•	
Reforestation							•
Protecting wetlands					•	•	
Rubbish collection			•			•	•
Industrial waste disposal facilities					•		
Infiltration maximization		•					
Retention ponds upstream		•					
Incentives to leave exposed areas						•	
Resettlement	•		•	•	•	•	
Dwellings built out of durable materials			•				
Community awareness			•	•	•		
Monitoring areas where construction is not permitted			•		•		
Public health			•				
Social insurance schemes		•			•		
Sea walls						•	

(IMWI 2013; Nguimalet 2007; Lewis 2009; Karley 2009; Commissioner for environment 2012; Chap. 13; République du Niger and Ville de Niamey 2012)

migration away from affected areas to cities, which results in an increase in the number of informal settlements in flood-prone city areas. Disaster risk assessment, land use planning, early warning, retrofitting, and formulation of emergency and reconstruction plans were among the first mitigating measures to be proposed (UN-Habitat 2007c). Cities are therefore seen as drivers of climate change through their energy consumption and the release of greenhouse gases. At the same time, the high concentration of poor people makes cities more vulnerable to climate change.

Table 2.8 Large cities south of the Sahara: local assessment tools, strategies, and plans for flood control

Assessment tools, strategies, and plans	Addis Ababa	Bangui	Cape Town	Cotonou	Dakar	Durban	Kampala	Lagos	Maputo	Niamey	Ouagadougou
Loss assessment					•						•
CC assessment							•			•	
Risk map									•		
Evaluation of needs								•			
Strategic agenda for CC adaptation	•										
CC adaptation strategy								•			
Climate adaptation action plan			•			•					
Storm water drainage master plan					•						
CC protection				•							
Local development plan adapted to CC										•	

IMWI 2013; Nguimalet 2007; Mukheibir and Ziervogel 2007; République du Sénégal 2012; Lewis 2009; UN-Habitat 2010b; Commissioner for environment 2012; République du Niger and Ville de Niamey 2012; Ville de Niamey 2012; République du Burkina Faso 2010 (see Chap. 12)

The year following publication of the *Report* saw the launch of the City and Climate Change Initiative in four cities, including Kampala and Maputo in Africa. The initiative was later extended to include Beira, Bobo Djoulasso, Kigali, Mombasa, Saint Louis, Vilankulo, and Walvis Bay (UN-Habitat 2012a).

In the subsequent 2009 *Global Report on Human Settlements*, climate change was recognized as the most important environmental issue (UN-Habitat 2009, p. 5). However, the understanding of climate change in urban environments remained generic in nature. Studies examining Kigali, Maputo, and Mombasa (UN-Habitat 2010) are far from having identified its salient characteristics.

These projects operate within a framework of international agreements that includes the Hyogo Framework for Action (or HFA), which prioritizes early warning systems and measures for reducing specific risks and is supported by the secretariat of the UNISDR (2007).

A system for identifying risk and setting up measures—although its main stages have been defined—remains rather difficult to implement. Mainstreaming flood risk reduction measures into the Millennium Development Goals and into poverty reduction strategies and infrastructure strategies is particularly time-consuming (IPU and UNISDR 2010).

A comparison of 11 large cities particularly exposed to flooding (Table 2.8) allows us to observe the adoption of a wide variety of assessment tools, strategies, and plans for flood risk reduction. Almost every city has some kind of instrument for evaluating damage and mapping risk and needs, and two-thirds of cities have planning tools. On the other hand, the adoption of local planning tools designed to reduce flood risk and facilitate adaptation is a complex process, as shown by the recent survey of promising practices in Africa, which consist entirely of single adaptation measures (UN-Habitat 2012b).

2.5 Conclusion

The connection between climate change and flooding in large cities south of the Sahara is not yet fully understood.

As regards recording and tracking rainfall patterns, a single weather station is not sufficient for vast urban areas such as Abidjan, Dar es Salaam or Maputo. Without multiple data points for a given city, the analysis of annual rainfall cannot be particularly meaningful. Calculation of rainfall recurrence intervals is absolutely essential if we wish to define the hazard (see Chaps. 9 and 12), but this is rarely done, and analysis of rainfall intensity and duration has been carried out in only a very few cases (De Paola et al. 2013). More precise information on daily rainfall (mm/hour) would allow us to understand whether rainfall has intensified over time. Such findings could point to climate change and would, in any case, be pertinent, particularly when planning the scale and location of adaptation measures.

The case studies considered demonstrate that flooding in riverside cities is not caused by local precipitation, but rather by precipitation affecting the entire watershed upstream of the city. This has two important implications: firstly, it is difficult to estimate the impact of future hazards; secondly, adaptation measures to reduce exposure and vulnerability cannot be limited to the urban area, but must also involve rural surroundings. The practical consequence of this observation is that adaptation measures for large cities should actually be adopted to some extent (perhaps to a great extent) outside of their administrative boundaries and agreed to together with other municipalities.

Adaptation assessment is still underdeveloped in Sub-Saharan Africa. Single flood risk reduction measures, including planning tools, are quite dissimilar to each other and do not follow the recommendations issued by the main multilateral organizations that deal with this matter.

The drainage canals in African cities are notoriously ineffective, and current rainfall levels are enough to cause persistent flooding. Resettlement of the entire exposed population exceeds the operating abilities of local governments. The only solution is therefore to ensure that what already exists works properly.

References

Adedeji OH et al (2012) Building capabilities for flood disaster and hazard preparedness and risk reduction in Nigeria: need for spatial planning and land management. J Sustain Dev in Africa 1(14):45–58

Adelekan IO (2000) A survey of rainstorms as weather hazards in southern Nigeria. Environmentalist 20:33–39

ADRC (2005) Total disaster risk management. Good practices, Asian disaster reduction center, Kobe

Aderogba K, Oredipe M, Oderinde S, Afelumo T (2012) Challenges of poor drainage system and floods in Lagos metropolis, Nigeria. Int J Soc Sci Educ 2(3):412–427

Angel S, Sheppard SC, Civco DL et al (2005) The dynamics of global urban expansion. World Bank, Washington

Blakie P, Cannon T et al. (1994) At risk: natural hazards, people's vulnerability and disasters. Routledge, London, pp 333–352

Bloch R, Jha AK, Lamond J (2012) Cities and flooding. A guide to integrated urban food risk management for 21st century. The World Bank, Washington

Braccio S, Tiepolo M (2013) Atlas des resources locales: Kébémer, Louga (Sénégal), Niamey (Niger). http://cooptriangulaireins.net/

Brinkmann J, von Teichman K (2010) Integrating disaster risk reduction and climate change adaptation: key challenges-scales, knowledge, and norms, sustainable science. Springer, Heidelberg

Commissioner for environment (2012) Towards a Lagos state climate change adaptation strategy

Conway G (2009) The science of climate change in Africa: impacts and adaptation, Grantham Institute for Climate Change. Discussion paper, 1

Conway D, Mould C, Bewket W (2004) Over one century of rainfall and temperature observations in Addis Ababa, Ethiopia. Int J Climatol 24:77–91

Crichton D (1999) The risk triangle. In: Ingleton J (ed) Natural disaster management. Tudor Rose, London

Da Costa H, Kandja KY (2002) La variabilité spatio-temporelle des precipitations au Sénégal depuis un siècle. In: Proceedings of the fourth international FRIEND conference, Cape Town

Davidson R (1997) An urban earthquake disaster risk index, the John A Blume earthquake engineering center, department of civil engineering, Report no. 121. Stanford University, Stanford

De Paola F, Giugni M, Topa ME, Coly A et al. (2013), Intensity-duration-frequency (IDF) rainfall curves, for data series and climate projection in African cities. Poster CLUVA project

De La Cruz-Reyna S (1996) Long term probabilistic analysis of future explosive eruptions. In Scarpa R and Tilling RI (eds) Monitoring and mitigation of volcano hazards, Springer, Berlin, pp 599–629

Dilley M, Chen RS, Deichmann U, Lerner-Lam AL, Arnold M (2005) Natural disaster hotspots. A global risk analysis, World Bank

Gotangco K, Perez R (2010) Understanding vulnerability ad risk: the CCA-DRM nexus. Klima Climate Change Center, Manila

Granger K, Jones T, Leiba M, Scott G (1999) Community risk in Cairns: a multi-hazart risk assessment. AGSO cities project Report no. 1

GTZ (2002) Disaster risk management. Working concept

Heat T, Parker A, Weatherhead K (2010) Impact of climate change on peri-urban areas in Lusaka. Report Cranfield University, London

IADB, Hahn H, Villagrán de León JC (2003) Comprehensive risk management by communities and local governments. Indicators and other disaster risk management instruments for communities and local governments

IATF CCDRR (2005) Disaster risk reduction tools and methods for climate change adaptation

IMWI (2013) Strategic agenda for adaptation to urban water mediated impacts of climate change in Addis Ababa. Ethiopia

IPCC (2007) Climate change 2007: synthesis report. An assessment of the intergovernmental panel on climate change

IPCC (2010) Managing the risks of extreme events and disasters in Africa. Lessons from the IPCC SREX report

IPCC (2012) Managing the risks of extreme events and disasters to advance climate change adaptation. Special report of the intergovernmental panel on climate change. Cambridge University Press, Cambridge

IPU, UNISDR (2010) Disaster risk reduction: an instrument for achieving the millennium development goals

Ishaya S, Ifatimehin OO, Abale IB (2009) Mapping flood vulnerable areas in a developing urban centre of Nigeria. J. Sustain Dev Africa 11(4):180–194

Jaiteh MS, Sarr B (2010) Climate change and development in Gambia. Challenges to eco system goods and services. http://www.columbia.edu/~msj42/pdfs/ClimateChangeDevelopment Gambia_small.pdf

Jankowska MM, Weeks JR, Engstrom R (2011) Do the most vulnerable people live in the worst slums? A spatial analysis of Accra. Ghana Annals GIS 17(4):221–235

Karley NK (2009) Flooding and physical planning in urban areas in west Africa: situational analysis of Accra, Ghana. CCASP TERUM 4(13):25–44

Kaynia AM, Uzielli M, Nadim F, Lacasse S (2008) A conceptual frame work for quantitative estimation of physical vulnerability to landslides. Eng Geol 102(3–4):251–256

Kolawole OM, Olayemi AB, Ajayi KT (2011) Managing flood in Nigerian cities: risk analysis and adaptation options—Ilorin City as a case study. Arch Appl Sci Res 2(1):17–24

Lewis M (2009) Climate change adaptation planning for a resilient city 2010/11, Durban municipality, environmental planning and climate protection department. http://www.durban.gov.za

Mahé G (2006) Variabilité hydro climatologique et envoronnementale en Afrique de l'Ouest et centrale : une approche des mécanismes à l'origine du changement des relations pluie-débit au cours du 20ème siècle

Mason SJ, Waylen PR, Mimmack GM, Rajaratnam B, Harrison JM (1999) Changes in extreme rainfall events in South Africa. Clim Change 41:249–257

Mitchell T, Van Aalst M (2008) Disaster risk reduction in climate change adaptation. A review for DFID, London

Mukheibir P, Ziervogel G (2007) Developing a municipal adaptation plan (MAP) for climate change: the city of Cape Town. Environ Urban 19(1):143–158

Nabegu AB (2009) Overview of climatic data availability for Kano region and suggestions for coping with climate change in the region. Techno Sci Afr J 3(1):100–107

Nguimalet CR (2007) Population et croissance spatiale : diagnostic et implications pour une gestion urbaine de Bangui (République centrafricaine), paper, PRIPODE Workshop, Nairobi

Okyere CY, Yacouba Y, Gilgenbach D (2012) The problem of annual occurrences of floods in Accra: an integration of hydrological, economic and political perspectives, ZEF Bonn. Universität Bonn, Bonn

Orok HI (2011) A GIS-based flood risk mapping of Kano city, Nigeria, thesis. University of East Anglia, Norwich

Rain D, Engstrom R, Ludlow C, Antos S (2011) Accra Ghana: a city vulnerable to flooding and drought-induced migration

République de Djibouti, MHUEAT, DATE (2001) Communication nationale initiale de la République de Djibouti à la convention cadre des Nations Unies sur les changements climatiques, December

République du Burkina Faso (2010) Inondations du 1er septembre 2009 au Burkina Faso. Evaluation des dommages, pertes et besoins de construction, de reonstruction et de relèvement. Evaluation conjointe. Rapport provisoire, Avril, Nations Unies-Banque Mondiale

République du Niger, Ville de Niamey (2012) Plan de développement communal. Version révisée

République du Sénégal (2010) Rapport d'évaluation des besoins post catastrophe. Inondations urbaines à Dakar 2009

République du Sénégal, ADM (2012) Etude du Plan directeur de drainage des eaux pluviales de la région périurbaine de Dakar, May

Saley MB, Tanoh R et al. (2009) Variabiltié spatio-temporelle de la pluviométrie et son impact sur les ressources en eaux souterraines: cas du district d'Abidjan (sud de la Côte d'Ivoire), paper presented at the 14e colloque International en évaluation environnementale, Niamey 26–29 mai

Tamru (2002) L'émergenge du risque d'inondation à Addis-Abeba: pertinence d'une étude des dynamiques urbane comme révélatrie d'un processus devulnérabilisation. Annales de Géographie, 627–628, 614–635

Turner BL et al. (2003) A framework for vulnerability analysis in sustainability science. Proc Nat Acad Sci 100(14):8074–8079

UNDHA (1992) Internationnally agreed glossary of basic terms related in disaster management, Geneva

UNDP (2004) Reducing disaster risk: a challenge for development. A global report, UNDP, New York

UNEP/OCHA Environment unit (2012) Keeping up with megatrends. The implication of climate change and urbanization for environmental preparedness and response

UN-Habitat (2004) Guidelines for reducing flood losses

UN-Habitat (2007c) Global report on human settlements 2007, vol. 3—Mitigating the impacts of disasters: policy, directions. Earthscan, London

UN-Habitat (2009) Global report on human settlements 2009. Planning sustainable cities. Earthscan, London

UN-Habitat (2010a) Planning for climate change. A strategic, values-based approach for urban planners

UN-Habitat (2010b) Climate change assessment for Kampala, Uganda. A summary

UN-Habitat (2011) Cities and climate change: policy directions. global report on human settlements 2011. Earthscan, London-Washington

UN-Habitat (2012a) Mid-term evaluation of the cities and climate change initiative, December

UN-Habitat (2012b) Promising practices on climate change in urban Sub-Saharan Africa

UN/ISDR (2002) Living with risk: a global review of disaster reduction initiatives, United Nations, Geneva

UNISDR (2007) Hyogo Framework for Action 2005-2015: Building the Resilience of Nations and Communities to Disasters. UNISDR, Geneva

UNISDR (2009) UNISDR terminology on disaster risk reduction. UNISDR, Geneva

UNITAR (2009) Overview of flood-affected sectors in Ouagadougou. Burkina Faso

USAID (2010) Risk assessment in cities

Villagrán de León JC (2004) Manual para la estimación cuantitativa de riesgos asociados a diversas amazenas, Acción Contra el Hambre, ECHO, CONRED, Villatek, Projectó de gestion local de desastres en lós municípios de Jocotán, Camotán y San Juan Ermita, Chiquimula

Ville de Niamey (2012) Rapport de la commission des affaires sociales culturelles et sportives, 25 aout

Wisner B, Blaikie P, Cannon T (2003) At risk. Natural hazards, people's vulnerability and disasters. Routledge, London

World Bank (2009) Framework for city climate change assessment. Buenos Aires, Dehli, Lagos, and New York

Part II
Urban Regions Under Climate Stress.
A Case Study: Dar es Salaam, Tanzania

Chapter 3
Climate Change Impacts and Institutional Response Capacity in Dar es Salaam, Tanzania

Dionis Rugai and Gabriel R. Kassenga

Abstract This chapter addresses causative factors and institutional response capacity to cope with the impacts of climate change in Dar es Salaam. Rapid population growth in the city has increased both environmental and socio-economic pressures, and has led to a worsening of settlement systems and patterns. Meanwhile, haphazard urban development has led to land degradation due to stone and sand mining, air and water pollution from untreated domestic and industrial waste, land and scenery pollution due to solid waste, the disappearance of green belts, and the loss of biodiversity. Recurrences of drought conditions and increased rainfall intensity have had significant social and environmental impacts, resulting in power and food shortages, losses of livestock and agricultural crops, and flood-related damages to infrastructure, human settlements, and livelihoods. Such conditions often correspond to the spread of diseases like malaria, diarrhea, and cholera, which have economic impacts for the government and families. Moreover, coastal erosion, loss of coastal and marine ecosystems, saline intrusion in freshwater bodies, inundation of low-lying coastal areas, and reduced freshwater flows due to sea level rise are evident. A number of adaptation strategies are in place that entail community involvement in planning processes.

Keywords Climate and weather hazards · Urban growth · Social and environmental impacts · Adaptation strategy · Dar es Salaam

D. Rugai (✉) · G. R. Kassenga
School of Environmental Science and Technology, Ardhi University, 35176, Dar es Salaam, Tanzania
e-mail: dionisr@gmail.com

G. R. Kassenga
e-mail: kassengagr@gmail.com

S. Macchi and M. Tiepolo (eds.), *Climate Change Vulnerability in Southern African Cities*, Springer Climate, DOI: 10.1007/978-3-319-00672-7_3,
© Springer International Publishing Switzerland 2014

3.1 Introduction

Dar es Salaam is located in the eastern part of the Tanzanian mainland, between latitudes 6°36′ and 7°0′ South and longitudes 39°0′ and 33°33′ East. It is bounded by the Indian Ocean on the east and the Coast Region on the other sides (Fig. 3.1). Dar es Salaam is the administrative and economic hub of Tanzania, although Dodoma Municipality has been designated as the national capital. The total surface area of Dar es Salaam is 1,800 km^2, with 1,393 km^2 of landmass, including eight offshore islands. The remaining 407 km^2 is covered by water bodies. In 2004, the built up area was about 698 km^2 (DCC 2004). Dar es Salaam covers about 0.19 % of the entire land area of mainland Tanzania (URT 2011).

Coastal shrubs, Miombo woodland, coastal swamps, and mangrove trees represent the main natural vegetation cover. The city is generally divided into four distinct landforms based on its morphological characteristics: (a) upland plateau, consisting of hills to the west and north, on average 100–200 m above mean sea level, with steep slopes as high as 330 m, and characterized by well drained unconsolidated gravely clay bound soils; (b) inland alluvial plains, which contain rivers flowing in from the Pugu Hills to the east, and Dar es Salaam Harbor, which snakes inland for nearly 10 km and is characterized by poorly drained silt clay soils enriched with organic matter; (c) coastal plains, characterized by overlain Pleistocene clay with relatively uniform relief, between 15 and 35 m above sea level and with a maximum slope of 3 % (continuing for 10 km west of the city, the plain is narrowest at Kawe—2 km—then expands to 8 km at the Mpiji River. The plain reaches a maximum width of 15 km in the southwest, where the relief becomes less uniform, gradually merging with land surround the mouth of the Mzinga River); and (d) shoreline and beach, which comprise the coastal area immediately adjacent to the sea. As the latter is characterized by sand dunes and tidal swamps, Dar es Salaam does not have extensive coastal lowlands (Kebede and Nicholls 2011: 7–8). The unique morphology of Dar es Salaam places the city in its own category in terms of both ecology and services as compared to other cities and towns in Tanzania.

The population of Dar es Salaam is estimated to be about 4,364,541 and the city is growing rapidly, at a rate of 5.6 % per year from 2002 to 2012 (NBS 2013).

The rates of urban growth and population increase have outpaced and compromised the local authority's capacity to provide basic services such as health, education, housing, water, and sanitation to its people. The infrastructure provided is inadequate, uncoordinated, and lags behind the pace of urban growth and other development dynamics (URT 2011). Rapid urban expansion has continued despite the paucity of resources and capacities to provide infrastructure and amenities. Most of the city's growth has occurred along the central and northern part of the coastline, with a great majority of the population living in unplanned and informal settlements (Kebede and Nicholls 2011). It is estimated that about 75 % of the population in Dar es Salaam live in over 100 unplanned or informal settlements occupying an area of 10,000 ha. Poor social and environmental conditions

Fig. 3.1 Location of Dar es Salaam City (reproduced from URT 2011)

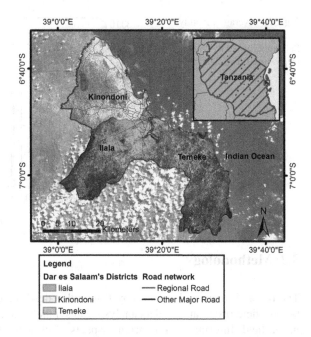

resulting from poor economic performance (especially low economic growth—4 % annually), and ill-conceived strategies to manage urbanization in poverty create favorable conditions for the formation of slums.

Dar es Salaam remains one of the most vulnerable cities in the country, particularly as regards climate change-induced hazards such as floods and sea level rise. This is especially so because more than 75 % of city inhabitants live in unplanned areas with little or no social services or other basic infrastructure such as roads, storm water drains, sewage systems, lighting, clean and potable water, health and education services (within the reach of affordable hospitals, clinics, dispensaries, and schools), police posts and stations, and fire and ambulance services (URT 2011). The city is vulnerable to climate-induced hazards such as floods, sea level rise, coastal erosion, water scarcity, and the outbreak of disease. The low-lying coastal areas of the city accommodate many people, and contain important ecosystem services and significant economic activities, such as port infrastructure, that are key to national and regional trade and import/exports; this could be threatened by future extreme climate conditions (Kebede and Nicholls 2011). The lack of public resources coupled with insensitivity, unawareness or lack of knowledge on climate change-induced hazards and threats, as well as the low prioritization such risks receive as result, further compound the problem.

Dar es Salaam, like many other coastal cities in Africa, is facing the challenge of adapting to the impacts associated with climate change and variability. The purpose of the present chapter is to analyze climate change in Dar es Salaam, illustrate climate change impacts, and identify the adaptation strategies employed by city residents and government institutions as well as the limitations thereof.

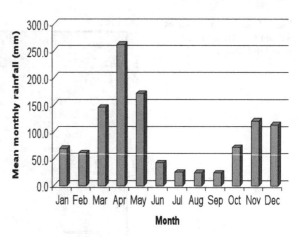

Fig. 3.2 Average monthly rainfall (mm) in Dar es Salaam City, 1960–2009 (reproduced from URT 2011)

3.2 Methodology

The research methodology employed included desk searches, stakeholder consultations, field observations, data analysis and synthesis, and discussions with experts in the field. Information on various aspects of climate change impacts and adaptation was gathered from secondary sources (literature, reports, and experience). Additional information was then collected through interviews and discussions with government officials, representatives of institutions, and non-governmental organizations.

3.3 Climate and Climate Change

3.3.1 Climate

The climate of the Dar es Salaam has the tropical characteristics of moist savannahs, with a dry season of 2.5–5 months (Dongus 2001). Usually, it has two rainy seasons: the long rainy season (March–May) with monthly average rainfall of 150–300 mm; and the short rainy season (October–December), with monthly average rainfall ranging from 75–100 mm. From 1960 to 2008, the annual rainfall range in the City was between 800 and 1300 mm (URT 2011), with an average annual rainfall of 1042 mm and considerably monthly variation, as shown in Fig. 3.2. Dar es Salaam experiences a bimodal rainfall regime. The short rains (*Vuli*) usually occur in October–December, while the long rains (*Masika*) occur in March–May (De Castro et al. 2004).

Generally, Dar es Salaam has a humid climate with average monthly temperatures that vary from 26 °C in August to 35 °C in December and January. With a typical coastal equatorial climate, Dar es Salaam experiences small seasonal and daily variations in temperature. The average monthly temperature ranges from a maximum of 31.5–32.1 °C to a minimum of 18.1–18.6 °C (De Castro et al. 2004). The mean daily temperature is about 26 °C, the mean seasonal range is about 4 °C, and the mean daily range is about 8 °C. Relative humidity reaches 100 % on almost every night of the year and rarely drops below 55 % during the day (URT 2011).

3.3.2 Climate Change

Global climate change trends are prominently reflected by projections for Tanzania's climate. For example, mean annual temperatures and average daily temperatures in Tanzania are expected to rise by between 2 and 4 °C by 2075 as a direct consequence of climate change (URT 2003). Recent Tanzanian climate change studies predict an increase in extreme weather events, particularly flooding, droughts, and cyclones, and more intense, frequent, and unpredictable tropical storms (URT 2003; Shemsanga et al. 2010).

Tanzania contributes relatively little to the causes of global warming compared to other countries. Its main contributions are deforestation, overgrazing, mining, air pollution from industries and vehicles, and land use changes (Shemsanga et al. 2010). In terms of contribution by sector, land use change contributes more to global warming than greenhouse gas emissions from fossil fuels, due to the country's low level of development. Dar es Salaam, and Tanzania in general, contributes significantly to carbon sequestration via its massive ocean environment, wetlands, forests, and land. Thus, efforts to combat climate variability in the country should focus on land use change.

Over the past few decades, rainfall in Dar es Salaam has been subject to high inter-annual variability (Fig. 3.3).

The rainfall trend in Fig. 3.3 shows that rainfall has decreased from about 1200 mm per year in the 1960s to about 1000 mm in 2010. For instance, amounts of both short and long rains (*Vuli* and *Masika*, respectively) were higher in the past and have decreased considerably in recent decades (Fig. 3.4). Both modalities of rainfall regime appear to have become less predictable, with increased seasonal shifts. For instance, the *Masika* now occur earlier than in the past, while the *Vuli* have been reduced to a near negligible amount in some years, and with their peak shifting from November to December (URT 2011). Although rainfall patterns reflect a downward trend, the average minimum rainfall values are still well above 600 mm (Fig. 3.3), implying that the risk of drought may not yet be critical. Analysis of daily and hourly rainfall data will be needed in order to ascertain the length of periods without rain during the rainy season, which will be important in more accurately determining the impact of climate change on rainfall patterns.

Although future rainfall patterns are uncertain, variability is likely to increase and intensification of heavy rainfall is expected. Thus, the impacts of flooding may become

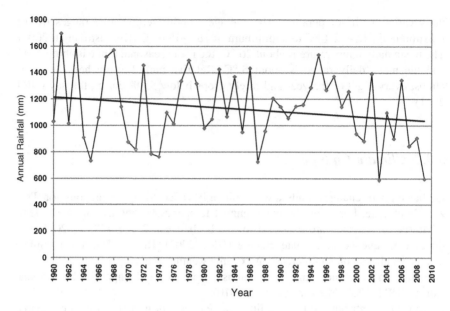

Fig. 3.3 Rainfall trends at Dar es Salaam International Airport in Dar es Salaam City, 1960–2009 (modified from URT 2011)

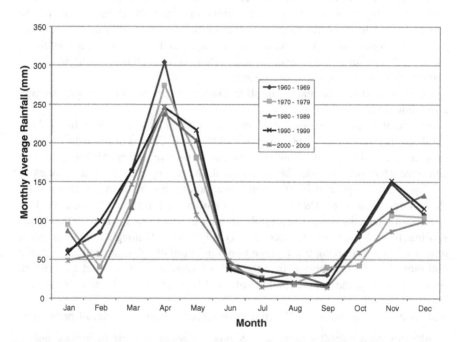

Fig. 3.4 Monthly rainfall in Dar es Salaam City, 1960–2009 (data from Tanzania Meteorological Agency, Dar es Salaam, 2009) (modified from URT 2011)

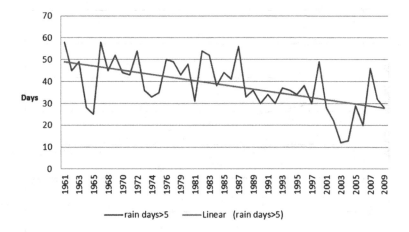

Fig. 3.5 Number of rain days per year at Wazo Hill (North of city center) (modified from START Secretariat 2011)

increasingly severe, particularly when combined with socio-economic projections, unless adaptation measures are implemented (START Secretariat 2011, p. 32).

It should be noted that the observed downward average rainfall trend portrayed by data from Dar es Salaam International Airport may not accurately represent the situation in the whole city. However, analysis of rainfall patterns in Dar es Salaam involving four locations representing the north, south, east, and west of the city center conducted by the START Secretariat (ibid) also showed a declining trend at all four locations, as summarized in Fig. 3.5. A rain day in Fig. 3.5 is defined as a day when at least 5 mm of rainfall were recorded. The decreasing rainfall pattern recorded specifically at Wazo Hill (north of city center) is also depicted in Fig. 3.6. These observations suggest that mean annual rainfall for the entirety of Dar es Salaam is declining.

Meteorological data indicate that the average annual minimum temperature steadily increased from 20 °C in the mid 1980s to 21 °C in 2000, and has continued to rise from 2000 to 2010 (URT 2011) (Fig. 3.7). The mean maximum temperature has also consistently increased over the last few decades, from 30.6 °C (in 1987) to 30.8 °C (in 2003), for an overall temperature increase of about 0.2 °C.

Available meteorological data show that from 1980 to 2008 there has been an increase in the minimum temperature by about 2 °C (Fig. 3.7). The period between 1999 and 2008 recorded the increased maximum temperatures for every month of the year (URT 2011). However, it should be pointed out that that rainfall and temperature records of short duration (<50 years) can be subject to bias due to interannual and decadal variability.

The combined risks of increases in mean temperature (Fig. 3.7) and fewer rainy days per year include lengthier dry seasons and intensified droughts. As evinced by

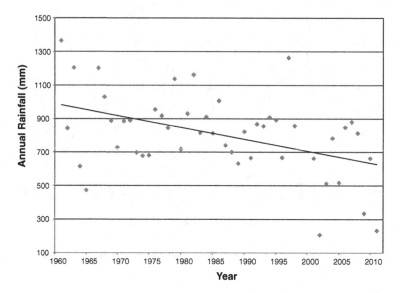

Fig. 3.6 Rainfall trend at Wazo Hill (north of city center) in Dar es Salaam City, 1960–2011 (plot data from Tanzania Meteorological Agency, Dar es Salaam)

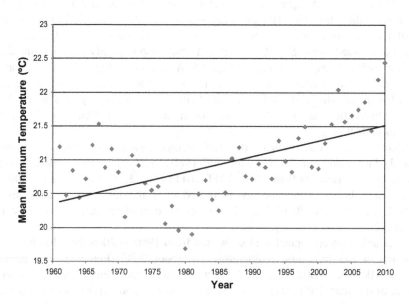

Fig. 3.7 Mean minimum temperature at Dar es Salaam International Airport 1961–2010 (plot data from Tanzania Meteorological Agency, Dar es Salaam)

the droughts of 2006 and 2008/2009, and the floods of 2009/2010, extreme climatic events can severely impact sectors such as agriculture, transport, energy and health, with adverse socioeconomic implications (START Secretariat 2011: p. 32–33).

3.4 Climate Change Impacts

Vulnerability (defined here as the propensity or predisposition to be adversely affected) to current climate conditions in Dar es Salaam is high, and is a function of socioeconomic factors (high population and overcrowding, poverty, malnutrition, exposure to disease), lack of adequate infrastructure and municipal services (e.g., waste removal, provision of clean water, access to working sanitation and drainage systems), poor hygienic practices, and climate variability (heavy rainfall and drought) (START Secretariat 2011). However, the main limitation of the study conducted by the START Secretariat (2011) is that vulnerability to climate conditions was not quantified.

Climate change in Dar es Salaam is strongly suspected to be responsible for seasonal variability and reduction of precipitation, temperature increase, sea level rise, drought, and floods. Identified climate change impacts include increased incidences of animal and human diseases, salinization of groundwater, and widespread damage to infrastructure, human settlements, livelihoods, and other property (URT 2003).

3.4.1 Floods

Dar es Salaam is among the many coastal areas that are experiencing flooding as one of the impacts of climate change. According to a recent study by Watkiss et al. (2011), 140,000 people in Dar es Salaam are currently below the elevation map's 10 m contour line, and over 30,000 are considered at risk. Long-term trends have suggested that the intensity and frequency of extreme heavy rainfall may increase in the wet seasons, which would imply greater flood risk (START Secretariat 2011). In December 2011, at least 23 people were killed in the worst floods to hit the Tanzanian capital in 50 years. Businesses were forced to close and thousands were left homeless as the city was inundated with floodwater (Fig. 3.8).

According to Casmiri (2009), areas prone to floods include Msasanibonde la mpunga (about 60 ha of mixed residential, commercial, and institutional settlements, and one of the fastest growing settlements in the Kinondoni municipality despite being prone to flooding), Msimbazi Valley, Jangwani (a slum area that floods during the rainy season almost every year), Mikocheni (where flooding is exacerbated by diversion of natural storm water drainage), and the city center (the most flooded area in the city, exacerbated by poor infiltration and an outdated,

Fig. 3.8 The flooded Msimbazi Valley in Dar es Salaam on 21st December 2011 (reproduced from http://teddykaegeles.blogspot.com/2011/12/dar-es-salaam-cityis-flooding.html accessed 15 April 2012)

non-functioning storm water drainage system). Often located in low-lying areas, shantytowns are highly prone to flooding caused by a variety of mechanisms, especially intense precipitation. It has been observed that heavy rainfall frequently causes flooding in these settlements, and, among other problems, contributes to increased incidence of disease. The historical frequency of flooding in various parts of Dar es Salaam is shown in Fig. 3.9. Low-lying areas with inadequate storm water drainage systems are more prone to flooding.

3.4.2 Sea Level Rise

Being a coastal city, Dar es Salaam is vulnerable to the impacts of sea level rise. However, available tidal gauge data, which cover the 2004–2011 period (7 years) may not be sufficient to draw meaningful conclusions regarding sea level rise in Dar es Salaam, though physical evidence does suggest that the problem exists. According to Kebede and Nicholls (2011), the major direct impacts of sea level rise include inundation of low-lying areas, loss of coastal wetlands, increased rates of shoreline erosion, saltwater intrusion and increased salinity in estuaries and coastal aquifers, and coastal flooding caused by higher water tables and higher extreme water levels. Already, coastal erosion is evident along many parts of the Dar es Salaam coastline, threatening public and private properties and affecting

Fig. 3.9 Flooding levels in Dar es Salaam (map prepared by Dionis Rugai in April 2012)

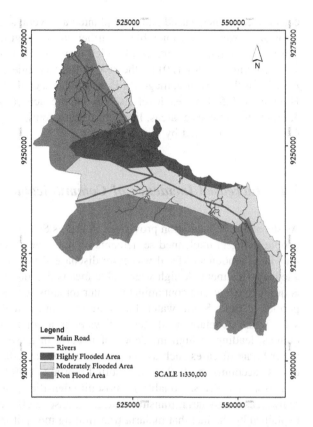

many beachfronts. At Kunduchi, north of the city, large areas have been lost to erosion, which now threatens other coastline infrastructure. Coastal erosion is evinced by collapsed trees, buildings, and other structures, including groins.

Coastal degradation and saltwater intrusion are major problems for Dar es Salaam's coastal areas today, and under projected climate change and possible sea level rise, coastal ecosystems would be highly threatened (Watkiss et al. 2011), affecting the livelihoods and ecosystem services of coastal communities. Residents of coastal wetlands affected by saltwater intrusion (such as the Suna, Mtoni Azimio, Msasani, and Bonde la Mpunga) have frequently needed to repair their houses as the saltwater is corroding the foundations and eating away cement bricks (START Secretariat 2011).

Low-lying coastal areas are susceptible to the effects of sea level rise, storm surges, and coastal erosion. For example, Kunduchi beach and Bahari beach in Dar es Salaam have been eroded to the point that heavy investments have had to be made to sustain them as beach areas. At Kunduchi, the headwater waves have advanced about 200 m in the past 50 years, as a result of which a mosque, five residential houses, and a historic fish market constructed in 1970s have been washed away or destroyed (Kebede and Nicholls 2011). Africana Hotel, constructed in 1967, is no longer operating due to damages. Sea walls have been constructed, stones placed and trees planted at several sites along the beach. These areas are particularly vulnerable to further coastal degradation, sea level rise, and storm surge intensification, all of which may worsen with climate change.

According to URT (2011), the risk to the coastline is imminent and it is estimated that the cost of damages in Dar es Salaam could be between 48–82 million USD for a 0.5–1 m sea level rise. Economic activities of importance to local communities in coastal areas, like salt making, tourism, and fishing are very likely to be further impacted by climate change.

3.4.3 Increased Incidence of Communicable Diseases

As stated above, sanitation provisions in Dar es Salaam are grossly deficient. Most people living in unplanned settlements do not have access to hygienic toilets and thus large amounts of fecal waste are discharged into to the environment without adequate treatment. A high water table means that during heavy rainfall, flooding is quick to occur and contaminated water remains stagnant in settlements for long periods of time. Storm water drains are frequently blocked by crude dumping of solid waste. Contaminated stagnant water is a common breeding site for mosquitoes, leading to high incidence of malaria in these settlements, and to other water borne diseases such as cholera, dysentery, and diarrhea. In Dar es Salaam, disease accounts for 48.9 % of reported mortality (URT 2003 and 2007).

Despite marked seasonality in mosquito densities, which usually peak with the rainy season, malaria transmission is intense and recurrent, which could be explained by the fact that malaria-transmitting mosquitoes are rampant throughout the year in Dar es Salaam due to frequent accumulation of water in residential environments. Life-threatening malaria occurs largely in children, commonly those under a year old (Schellenberg et al. 1999 in URT 2011). De Castro et al. (2004) report that from 2 to 10 % of school children living in urban Dar es Salaam are infected with malaria. Anemia, largely caused by malaria, is also common in adults and children.

Increases in mean minimum temperature (Fig. 3.4) are critical to the ecological ranges and survival of various organisms, including pathogens, pests, and disease vectors. For instance, survival of adult anopheles mosquitoes, the vector for malaria parasites, is usually influenced by variables like temperature, humidity, and rainfall (URT 2011).

Cholera is another health risk that is very sensitive to climate change. Cholera outbreaks in Dar es Salaam are usually associated with warm and wet seasons (Fig. 3.10); therefore any increase in the amount of rainfall coupled with increasing temperatures would add to cases of cholera. Coastal areas, including those in Dar es Salaam, are increasingly subject to cholera outbreaks caused by extreme weather events. In economic terms, the increased incidence of malaria and cholera in Dar es Salaam adds to the medical treatment costs born by individual households, communities, and the country as whole.

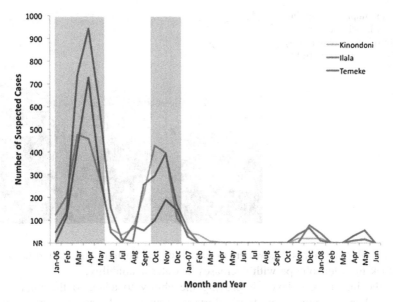

Fig. 3.10 Number of suspected cholera cases by month, year, and municipality in Dar es Salaam, 2006–2008 (reproduced from Penrose et al. 2010)

3.5 Climate Change Adaptation

Communities utilize and/or integrate their environmental knowledge into adaptation plans (URT 2011). For instance, shortages of piped water for domestic use, which are often attributed to climate change, have prompted increased use of water tanks (plastic or concrete) to harvest and reserve water in times of scarcity. Many city residents are also digging deep and shallow wells as sources of water for domestic uses in areas where older wells have been affected by saltwater intrusion, including in the Vikuruti area of Dar es Salaam. The use of such adaptive mechanisms depends considerably on the economic status of a given household, which may not be able to afford such facilities. In the absence of water from rainfall, urban farmers in Dar es Salaam have in many instances practiced small-scale irrigation, sometimes using untreated wastewater. Together with sea level rise, strong waves are accelerating beach erosion. This has necessitated artificial placement of sand on the beaches to create a coastal drain beach management system. Examples include groins to protect beach hotels north of Dar es Salaam (Fig. 3.11), and sea walls along Ocean Road beach.

There is also increased use of insecticide-treated bed nets (ITNs) to prevent contraction of malaria. This adaptive measure has been heavily promoted by the Government (URT 2011). Ricci (2011) reports that many residents in peri-urban areas of Dar es Salaam have also changed crop systems (e.g. moving from rice to

Fig. 3.11 Improvised groins
at Mbezi Beach area Dar es
Salaam (photograph taken by
Dionis Rugai in June 2011)

cassava, which requires less water) or have stopped farming and started breeding livestock in order to cope with decreases in water availability.

At the institutional level, improving the ability to adapt to the impacts of climate change requires long-term efforts to strengthen the adaptive capacity of each community. Tanzania has taken steps towards addressing the issue of climate change in its widest sense. It ratified the United Nations Framework Convention on Climate Change (UNFCCC) in 1996, and the government subsequently formulated a National Adaptation Programme of Action (NAPA). Among other objectives, the NAPA aims at improving public awareness of the impacts of climate change and potential adaptation measures, and at identifying and developing immediate and urgently needed actions to adapt to climate change and climate variability (URT 2007). Thus, the NAPA is an important step in the long process of helping national stakeholder groups understand the problem of climate change and the roles they can play in building resilience to potential impacts. The NAPA is expected to facilitate integration of adaptation issues into the development process, in order to address urgent and immediate needs for adaptation to the adverse impacts of climate change. Tanzania recently formulated the National Climate Change Strategy (2012) with the overall aim of enhancing technical, institutional, and individual capacities to address the impacts of climate change. The Strategy covers adaptation, mitigation, and cross-cutting interventions that are expected to enable Tanzania to benefit from the opportunities available to developing countries in their efforts to tackle climate change (URT 2012). The strategy seems to be ambitious as regards enhancement of various key stakeholders' capacities to deal with salient issues pertaining to climate change adaptation. However, it is too early to judge the effectiveness of this strategy in addressing the impacts of climate change.

In Dar es Salaam, there are only two notable programs, the Community Infrastructural Upgrading Program (CIUP) and the Strategic Urban Development Plan (SUDP), both of which target infrastructure improvements in poor areas, including roads, storm water drainage channels, and solid and liquid waste management facilities. The SUDP and the CIUP also seek to nurture community-based initiatives. Such initiatives need to be supported and expanded and, importantly, their benefits sustained. Another activity, which is normally carried out in an ad hoc fashion is the planting of trees along the beach, roadsides, near houses and in open spaces. Dar es Salaam also has a Disaster Management Unit that coordinates responses to disasters, including those related to climate change (URT 2011). Among the climate change-related disaster management responses are measures taken to prevent or minimize cholera outbreaks. However, preparedness for other extreme weather events, such as flooding and drought, is inadequate, as evinced by the recurrent flooding of the Msimbazi River Valley, where many valley residents are affected nearly every time floods occur. Such areas would be better left as greenbelts rather than human habitations. However, the main challenge of maintaining greenbelt areas is the prevention of human activities, especially the construction of houses, sand mining, and farming. The local government authorities (LGAs) have failed to prevent further construction of houses as well as to evict people living in Msimbazi Valley due to lack of resources and interference from leaders for political reasons.

3.6 Governance and Institutional Capacity for Climate Change Induced Hazards

A number of governance challenges that are closely related to climate change induced hazards have been identified in Dar es Salaam. Such challenges include weak urban planning and management systems, institutional practices that are unresponsive to climate change induced concerns and threats, the absence of specific policies and regulations for addressing climate change and related hazards, low capacity and awareness among grassroots (community) organizations, apathy among politicians and bureaucrats as regards climate change, and reluctance to embrace research findings in critical decision making issues.

Broadly speaking, LGAs and other organizations have some bearing in three important ways on the extent to which climate hazards affect the local community (Kates 2000): they influence how households are affected by climate impacts; they shape the ability of households to respond to climate impacts and pursue different adaptation practices; and they mediate the flow of external interventions in the context of adaptation. The capacity of particular institutions is important with respect to how they affect adaptation. But equally important are the linkages and interconnections they have with each other and peri-urban households; these affect the flow of resources and decision making power among social groups, and thus their capacity to adapt. The capacity of LGAs in Dar es Salaam in dealing with the impacts of climate change has been observed to be limited. The issues pertinent to the institutional capacity of LGAs in Dar es Salaam to deal with climate change matters and implementation of the NAPA are as follows:

1. Awareness of climate change issues among LGA officials and the general public is lacking.
2. The city's policies and programs for addressing current vulnerability and promoting adaptation to climate change are limited.
3. Mainstreaming of climate change and disaster risk management issues in development plans, strategies, programs, projects, and routine activities has yet to be done.
4. Members of city staff and municipal councils have a limited capacity to effectively analyze the potential impacts of climate change and to develop viable adaptation measures.
5. Climate change issues are not sufficiently addressed in policies and programs, since most were prepared before the late 1990s.
6. City planning departments' capacities to assess the long-term sectorial impacts of climate change are limited.
7. There is widespread apathy and insensitivity to climate change issues.
8. Views on the best approach for addressing climate change vary.
9. Coordination among experts in the municipalities is limited (the offices of the Town Planner, Engineer, and Health Officer tend to work independently of each other).
10. The capacity for investing in adaptation activities (protecting vulnerable populations, infrastructure, and economies) is still low due to financial constraints.

3.7 Conclusions and Recommendations

Due to the lack of long-term meteorological and tidal gauge data and other pertinent geological and topographical information, climate change vulnerability in Dar es Salaam cannot be assessed with reasonable accuracy. Notwithstanding these limitations, it can be generally observed that climate change will continue to exacerbate the problems experienced by Dar es Salaam's residents in areas such as health, living conditions, livelihoods, and economic security. Dar es Salaam is not adequately adapted to the current climate and has a large existing adaptation deficit that requires urgent action.

Urban development and poverty reduction policies and programs for the city need to be adjusted to not only better integrate disaster risk management approaches but to assume a long-term perspective that takes climate change into account. At the same time, existing policies and regulations that can reduce the vulnerability of the poor need to be better enforced. The capacity of Dar es Salaam's private sector entities needs to be better developed to enable them to engage actively in supporting adaptation and mitigation while also profiting from such endeavors. The government, in collaboration with city and municipal authorities, should undertake awareness raising and sensitization programs regarding the impacts of climate change on city residents and potential adaptation and mitigation options. Most importantly, there is an urgent need to build capacity, through mechanisms, institutions, and governance systems for LGAs and other institutions in Dar es Salaam, in order to enable them to help city residents adapt to the impacts of climate change.

References

Casmiri D (2009) Vulnerability of Dar es Salaam city to impacts of climate change. IIED, London

Dar es Salaam City Council (DCC) (2004) Dar es Salaam city profile. Tycooly Publishing Company, Dar es Salaam

De Castro MC, Yamagata Y, Mtasiwa D, Tanner M, Utzinger J, Keiser J, Singer BH (2004) Integrated urban malaria control: a case study in Dar es Salaam, Tanzania. Am J Trop Med Hyg 71(Suppl 2):103–117

Dongus S (2001) Urban vegetable production in Dar es Salaam (Tanzania)—GIS-supported analysis of spatial changes 1992–1999. APT-Reports No. 12, Freiburg

Kates R (2000) Cautionary tales: adaptation and the global poor. Clim Change 45(1):5–17

Kebede AS, Nicholls RJ (2011) Population and assets exposure to coastal flooding in Dar es Salaam (Tanzania): vulnerability to climate extremes. Global Climate Adaptation Partnership (GCAP), Southampton

National Bureau of Statistics (NBS) (2013) Tanzania population and housing census 2012. United Republic of Tanzania, Dar es Salaam

Penrose K, CastroMCd Werema J, Ryan ET (2010) Informal urban settlements and cholera risk in Dar es Salaam Tanzania. PLoS Neglected Trop Dis 4(3):e631. doi:10.1371/journal.pntd.0000631

Ricci L (2011) Peri-Urban livelihood and adaptive capacity: urban development in Dar Es Salaam. J Sustain Dev 7(1/2012):46–63

Shemsanga C, Omambia AN, Gu Y (2010) The cost of climate change in Tanzania: impacts and adaptations. J Am Sci 6(3):182–196

START Secretariat (2011) Urban poverty & climate change in Dar es Salaam, Tanzania: a case study. Final report prepared/contributed by the Pan-African START Secretariat, International START Secretariat, Meteorological Agency and Ardhi University Dar es Salaam. http://start.org/download/2011/dar-case-study.pdf. Accessed 12 June 2013

United Republic of Tanzania (URT) (2003) Initial National Communication under the United Nations Framework Convention on Climate Change (UNFCCC). Vice President's Office, Dar es Salaam

United Republic of Tanzania (URT) (2007) National Adaptation Programme of Action. The Vice President's Office, Division of Environment, Dar es Salaam

United Republic of Tanzania (URT) (2011) The Dar es Salaam City Environment Outlook. The Vice President's Office, Division of Environment, Dar es Salaam

United Republic of Tanzania (URT) (2012) National Climate Change Strategy. The Vice President's Office, Division of Environment, Dar es Salaam

Watkiss P, Downing T, Dyszynski J, Pye S et al (2011) The economics of climate change in the United Republic of Tanzania. Report to development partners group and the UK department for international development. http://economics-of-cc-in-tanzania.org. Accessed Jan 2011

Chapter 4
Climate Change Effects on Seawater Intrusion in Coastal Dar es Salaam: Developing Exposure Scenarios for Vulnerability Assessment

Giuseppe Faldi and Matteo Rossi

Abstract This chapter is part of a project that aims to identify climate change vulnerability scenarios for the inhabitants of Dar es Salaam's coastal areas, with specific reference to the phenomenon of seawater intrusion. The rapid urbanization that has taken place in Dar es Salaam over the past 20 years has resulted in a significant increase in anthropogenic pressure (qualitative and quantitative) on the coastal aquifer, causing an acceleration in groundwater salinization processes, due to both seawater intrusion and the leakage of pollutants. Seawater intrusion could be further amplified in the medium and long term by the expected consequences of climate change. This work presents the conceptual and methodological framework adopted to identify scenarios of population exposure to seawater intrusion, based on specific indicators of the phenomenon and their correlations with climate change representative variables. Based on historical data integrated with a specific hydrogeological survey campaign carried out in 2010, a first application of that framework has been tested on a limited portion of Dar es Salaam's coastal area. This allowed for a preliminary assessment of seawater intrusion indicators, as well as an evaluation of their efficacy when analyzing the evolution of the phenomenon over the last decade.

Keywords Biophysical system sensitivity · Vulnerability assessment · Saltwater intrusion · Coastal aquifer monitoring · Dar es Salaam

G. Faldi (✉)
Department of Astronautical, Electrical and Energetic Engineering, Sapienza University of Rome, Via Eudossiana 18, 00184 Rome, Italy
e-mail: giuseppe.faldi@yahoo.com

M. Rossi
Department of Civil, Building and Environmental Engineering, Sapienza University of Rome, Via Eudossiana 18, 00184 Rome, Italy
e-mail: matteo.rossi@uniroma1.it

S. Macchi and M. Tiepolo (eds.), *Climate Change Vulnerability in Southern African Cities*, Springer Climate, DOI: 10.1007/978-3-319-00672-7_4,
© Springer International Publishing Switzerland 2014

4.1 Introduction

The purpose of a vulnerability assessment in a climate change related study is to develop a qualitative understanding of the environmental, social, and economic processes that can turn the consequences of climate change into possible risk factors for communities (Downing and Patwardhan 2004; Füssel and Klein 2006). Vulnerability assessments can identify the regions and social groups that are most susceptible to the consequences of a particular disturbance, consequences that could be increased and aggravated by the effects of climate change, and can provide a knowledge base for the implementation of adaptation strategies for specific socio-economic contexts (Downing and Patwardhan 2004; IPCC 2007).

The methodology illustrated in this chapter aims to identify climate change vulnerability scenarios for the inhabitants of Dar es Salaam's coastal areas, with specific reference to the phenomenon of seawater intrusion.

Groundwater is the largest reserve of freshwater available worldwide, and thus plays a crucial role in the adaptability of the world population to the effects of climate change on rainfall, soil moisture content, and surface water (Margat 2006). Yet the link between surface water and many terrestrial and freshwater ecosystems is often overlooked. Since many hydrogeological systems are hydraulically interconnected with surface waters, a change in the regime of surface waters will certainly have direct consequences for the groundwater system. Therefore, underestimating the role of groundwater in water resource management could lead to inappropriate strategies with important consequences for the population and environment.

Recent IPCC assessment reports have concluded that very little is known about the relationship between groundwater and climate change (IPCC 2001, 2007, 2008). However, recent years have seen a constant increase in the scientific papers on the cause-effect relationships between groundwater and climate variability on a local and global scale (Eckhardt and Ulbrich 2003; Arnell et al. 2004; Barnett et al. 2004; Brouyère et al. 2004; Scibek and Allen 2006; Kundzewicz et al. 2008). The impact of climate change on groundwater is less direct when compared with surface water, mainly as regards the time scale of effects. The groundwater dynamic has much slower rhythms than the surface hydraulic system; therefore, while low or intense rainfall effects have direct and rapid consequences for the rivers regime, only after prolonged droughts will a decreasing trend be reflected by aquifer levels. Climate change usually acts as an effects multiplier in already altered hydrogeological systems, with obvious consequences for dependent eco-systems and communities (Appleton 2003). Besides increases in the frequency and intensity of extreme events and the expected rise in average sea level, other unpredictable trends must be taken into consideration. The complexity of the parameters and hydrogeological systems involved makes it difficult to even imagine the evolution of certain phenomena involving groundwater. Since high rainfall intensity leads to rapid soil saturation and a decrease in direct aquifer recharge due to infiltration, the annual trend variation in the rainfall regime due to climate change can adversely affect groundwater in the wetlands. By contrast, in arid and semiarid areas, only a very intense rainfall can counterbalance the high temperatures and high water evaporation rates affecting the first layers of soil. Under such conditions, the phreatic aquifers are in fact mainly recharged by water during flood events (Burns 2002). The impacts of climate change can also lead to a decrease in groundwater quality, as in the case of seawater intrusion in coastal aquifers (Kundzewicz et al. 2008). Phenomena such as freshwater supply loss due to a decrease in groundwater direct recharge, coastal erosion due to mean sea level rise, and increased groundwater withdrawals as an alternative water supply where surface waters are already in crisis can alter the saltwater/freshwater equilibrium, seriously affecting water resources and related activities for decades. Ataie-Ashtiani et al. (1999) have reported the effects of tide excursion on confined coastal aquifers, concluding that, especially in cases of large oscillations due to strong tides or mean sea level rise, an amplification of seawater intrusion can be observed, with serious consequences as regards transition zone thickness. Bobba et al. (2000) and Ranjan et al. (2006) have studied the effects of climate change on direct rainfall recharge and mean sea level rise, adopting numerical modeling as a tool to monitor the freshwater/saltwater interface. Sherif and Singh (1999) have reported on two case studies, in coastal areas of Egypt and India, that assessed the consequences of climate change for groundwater in terms of mean sea level rise. More generally, the effect of climate change on groundwater, and specifically on the phenomenon of seawater intrusion in coastal aquifers, seems today to be a subject of fervent scientific interest, as has been evinced by multiple sessions of international conferences dedicated to the topic[1] and by specific focus chapters of international associations and organizations.[2]

Regardless of its possible relationships with climate change, groundwater salinization is already a widespread problem that is conditioning the socio-economic development of many African coastal cities, such as Dakar in Senegal (Faye et al. 2004; Re et al. 2011), Accra in Ghana (Kortatsi and Jørgensen 2001; Kortatsi 2006), Lagos in Nigeria (Oteri and Atolagbe 2003; Adepelumi 2008), and Dar es Salaam in Tanzania (Mtoni et al. 2012), all of which are characterized by an extremely high urbanization rate and widespread practice of urban agriculture. In southern countries in particular, the rapid urbanization of coastal areas could encourage overexploitation of groundwater and an increase in sources of pollution (Mato 2002; Steyl and Dennis 2010), both of which cause deterioration of aquifer quality and increases in seawater intrusion.

Over the past 20 years, the population of Dar es Salaam has increased considerably (from 1.8 in 1992 to over 4.3 million in 2012, according to the national census), leading to unplanned development of the urban fabric (extended

---

[1] The International Hydrological Programme (IHP) of UNESCO and the International Association of Hydrological Sciences Symposium, *Water Quality: Current Trends and Expected Climate Change Impacts*. Proceedings available at: http://iahs.info/redbooks/348.htm.

[2] The International Association of Hydrological Commission on Groundwater and Climate Change: available at http://www.iah.org/gwclimate/.

peri-urban areas), the proliferation of informal settlements, and the deterioration of basic public services, together with a significant increase in the water demands of inhabitants (UN-HABITAT 2009). The inadequacy of the municipal water system, which is based only on surface water use, has led in recent years to massive groundwater exploitation in order to meet the growing water demand for anthropic activities (Kjellen 2006; URT 2011). Over the past 15 years, the number of wells has increased significantly, up from a few dozen to more than 2200 official private wells and an unknown number of informal boreholes (Mjemah et al. 2009; JICA 2012). Moreover, these numbers continue to increase (Mtoni et al. 2012). The increase in groundwater exploitation raises many issues of great scientific interest as regards the quantity, but also the quality of water resources, which are increasingly affected by salinization processes caused by anthropogenic pollution (Mato 2002; Mjemah 2007), a rise in saline fossil water (Mjemah 2007) and increasing seawater intrusion (Mtoni et al. 2012).

This work presents the conceptual and methodological framework adopted to identify scenarios of population exposure to seawater intrusion, based on specific indicators of the phenomenon and their correlations with variables representative of climate change (mean sea level, value of direct rainfall recharge, etc.). As described below, that framework has been applied for the first time to a limited portion of the Dar coastal area, using data from the literature integrated with a specific monitoring campaign carried out in 2010 and involving 21 boreholes; this allowed for a primary assessment of the seawater intrusion indicators, as well as an evaluation of their efficacy when analyzing the evolution of the phenomenon over the last decade.

4.2 Methodology

The present description of the methodology adopts the official climate change terminology (IPCC, UNFCCC), with reference to Füssel (2007) (Fig. 4.1). More specifically, it is assumed that vulnerability to climate change is

> the degree to which a system is susceptible to, or unable to cope with, the adverse effects of climate change, including climate variability and extremes. Vulnerability is a function of the character, magnitude, and rate of climate variation to which a system is exposed, its sensitivity and its adaptive capacity (IPCC 2007).

Vulnerability = f (Exposure, Sensitivity, Adaptive Capacity)

According to this approach, the vulnerability of individuals or communities is the result of the interaction of physical and socio-economic factors. Physical factors represent the potential that an environmental system be damaged by the consequences of a harmful event, thus identifying the degree of human exposure to disruption (connection between livelihoods and ecosystems). Socio-economic factors, in turn, represent the ability of individuals and communities to cope with

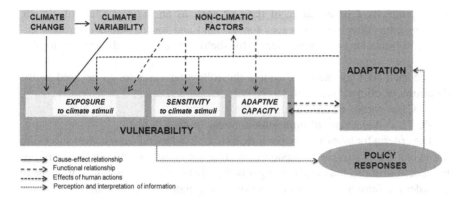

Fig. 4.1 General conceptual framework for vulnerability assessment (Modified from Füssel and Klein 2006)

the disturbance, absorb the impact, recover or adapt to change (Clark et al. 1998; Füssel and Klein 2006).

In this study, the analysis methodology focuses first on identifying various seawater intrusion scenarios for the Dar es Salaam coastal aquifer, and then on assessing the population's level of exposure to this disruption. The consequent definition of sensitivity and, therefore, the development of future vulnerability scenarios according to different adaptation options, will be the subject of future project activities. In order to apply this approach to studying seawater intrusion, it is necessary to identify the relationships, analytical or qualitative, between specific indicators related to the phenomenon (IFs) and the environmental variables related to climate change (VCCs). For global and local climate scenarios, those suggested by international reference organizations will be adopted (Ziervogel et al. 2003; Arnell et al. 2004; Lu 2006; IPCC 2007). Assuming that the evolution of seawater intrusion can usually be studied by following the advance of saltwater towards the inland, the analysis, monitoring, and evaluation will be conducted separately on different strips parallel to the coastline that traverse the whole study area.

Since the assessment of direct relationships and correlations between climate change and groundwater dynamics is a difficult task, identification of the VCCs influencing seawater intrusion must therefore be based on several general and scientifically well-founded assumptions:

- the average sea level rise will generally cause an acceleration of the advance of brackish water toward the inland;
- a significant freshwater discharge variation will cause a disruption of the natural equilibrium with the seawater near the coast;
- rainfall infiltration is directly related to precipitation and temperature patterns, and consequently to variations in their trends caused by climate change;

- the progressive urbanization of the Dar area, as an *indirect* effect of climate change, has direct consequences for the land and for variations in ground impermeabilization, and consequently for the decreasing rainfall infiltration rate.

The methodology therefore adopts the following variables to be used in the definition of climate scenarios:

VCC1 = mean sea level (m);

VCC2 = direct rainfall infiltration amount (mm).

The criteria for assessing the parameters to be used as *indicators* when surveying the evolution of seawater intrusion over time are closely related to the representativeness of those indicators as regards the phenomenon; it is also important to consider the feasibility of data collection during monitoring campaigns.

The following indicators were thus identified:

IF1 = average area electric conductivity (μS/cm);

IF2 = maximum area piezometric charge (m.a.s.l.);

IF3 = maximum 2000 μs/cm EC isoline distance from the coastline (brackish water limit).

Electric conductivity (EC) is directly related to groundwater salinity, while the piezometric survey in a given area may provide indications for the assessment of seawater intrusion hydrodynamics as well as their temporal evolution.

The methodology is based on an intense IF monitoring activity over multiple time periods, and on simultaneous VCC evaluation in order to identify the specific correlations. Therefore, for each time step, a set of measured IF and estimated VCC are used to assess, by interpolation, the existing analytical or qualitative relationships. A diagram of the proposed methodology is shown in Fig. 4.2.

4.3 Preliminary Application in the Dar es Salaam Coastal Area

4.3.1 Seawater Intrusion in Dar es Salaam

Seawater intrusion is the movement, transient or stationary, of seawater into a coastal aquifer. Mixing with aquifer groundwater, it creates a *transition* brackish water located between the freshwater and the seawater below (Bear et al. 1999; Araguás et al. 2004). Therefore, this phenomenon is normally ruled by the natural balance between the groundwater flow toward the sea and the saltwater flow toward the coast, depending on seasonal climate fluctuations and tidal excursions. When external disturbances alter the natural equilibrium (overexploitation of the aquifer, soil waterproofing, coastal erosion, sea level rise), the thickness of the transition zone in the hydrogeological system grows, due to salt migration from areas with greater concentration values to areas with lesser ones, until the new equilibrium is reached (Bear et al. 1999; Araguás et al. 2004).

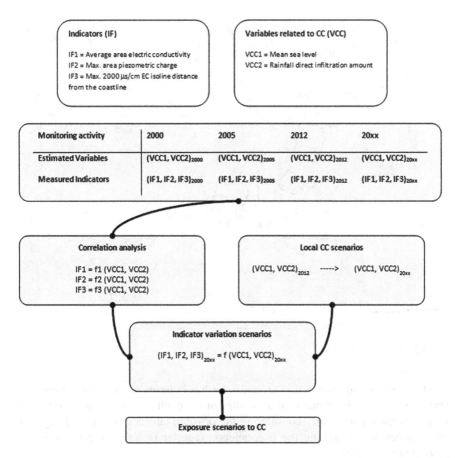

Fig. 4.2 Conceptual framework of the methodology adopted for assessing exposure to seawater intrusion

Although the lack of available hydrogeological and geochemical historical data did not allow for the development of detailed investigations (i.e. transition zone temporal evolution and spatial distribution assessment), some authors (Mtoni et al. 2012) have noticed an increase in seawater intrusion in recent decades, demonstrating the extent to which this phenomenon could be a possible threat to groundwater resources in Dar es Salaam. Some coastal areas in Dar, such as Msasani, Oysterbay, Masaki, Kigamboni, and Gerazani show a noticeable increase in groundwater salinity with respect to the past, which is an effect of the rising salt wedge (Mtoni et al. 2012). The seawater intrusion increase in recent years seems to be related to both the effects of overexploitation of groundwater, and of soil waterproofing due to urbanization, which have severe consequences for the direct recharge of the aquifer. This is evinced by the decrease in piezometric levels in the coastal area surveyed in recent studies, a possible cause of increases in the freshwater/saltwater interface.

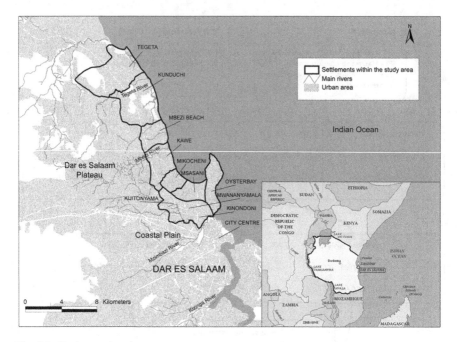

Fig. 4.3 Study area location

The growing importance that groundwater has in terms of the water supply for Dar es Salaam's inhabitants renders studies of the evolution of seawater intrusion very important, particularly considering the possible effects of climate change on the parameters determining the phenomenon (aquifer recharge, mean sea level, piezometric surface).

4.3.2 Study Area Characteristics

The study area (Fig. 4.3) has a surface of approximately 100 km^2, which extends along a 29 km stretch of coastline to the north of the city center and is bordered to the east by the Indian Ocean. The western boundary is the Dar es Salaam Plateau, which rises west of the Ocean along the entire study area. The southern boundary corresponds to the last segment of the Msimbazi River, which runs through the coastal plains before flowing into the Indian Ocean to the south of the Msasani Peninsula, while the northern boundary is located just north of the Tegeta River.

Analysis of saltwater intrusion in the Dar es Salaam coastal aquifer began with a reconstruction of the geological and hydrogeological structure of the coastal physical system within the study area.

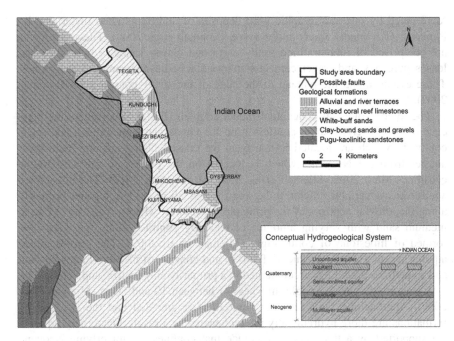

Fig. 4.4 Outcropping formations with the conceptual hydrogeological system

As for geological characteristics, the outcropping formations belong to two principle geologic periods: the Quaternary and the Neogene (Fig. 4.4). Formations from the Quaternary are found primarily in the areas near the coast, have a maximum thickness of about 150 m, and fill the tectonic depression, which originated in the lowering of formations from the Neogene (Mjemah 2007). Within the study area, the Quaternary is evidenced by three different stratigraphic units with heterogeneous characteristics: alluvial deposits (Pleistocene to Recent, *Alluvial and river terraces*), sandy deposits in the coastal plains (Pleistocene, *White-buff sands and gravels*) and limestone deposits (Pleistocene, *Raised coral reef limestone*) (Mjemah et al. 2009). The alluvial deposits consist essentially of sand, clay, and sometimes gravel, and they are found predominantly along the recent alluvial plains of the principle waterways (Mbezi River and Msimbazi River) and in the Kunduchi lagoon. The mostly sandy deposits consist of sand and white washed gravel, interspersed with clay lenses, and they appear next to the coastal plain within the study area. The limestone deposits consist of coral materials and are located primarily in a narrow strip that widens along the coast. They are generally eroded and covered with white-buff sand, except in the Msasani Peninsula and to the north of Kunduchi, where they appear at the surface (Mjemah 2007). Formations from the Neogene are spread predominantly throughout the central part of the Dar es Salaam bay, on the edge of the sandy coastal plains of the Quaternary, and have a thickness of up to 1000 m (Kent et al. 1971). Within the study area, the Neogene is represented by one stratigraphic unit: deposits of undifferentiated material (Mio Pliocene, *Clay-bound sands and gravels*) (Mjemah et al. 2009). These sediments are comprised of interstratified sandy clays and clayey sands, a few pure sand lenses, pure clay, and gravel, cemented in irregular bodies to form weak sandstones. These extend along the plateau to the edge of the sandy coastal plain (Mjemah 2007).

The hydrogeological conceptual model adopted (Fig. 4.4) considers the presence of two different sandy aquifers from the Quaternary (unconfined and semi-confined), which are located primarily in the coastal plain, and one multi-layer aquifer from the Neogene, which is located on the plateau to the west of the coastline (Mato 2002; Mjemah et al. 2009; Mtoni et al. 2012). The unconfined aquifer has a variable thickness between 5 and 50 m, and is located primarily along the alluvial deposits of the principle waterways, where it consists of fine to medium sand with varying amounts of clay. By contrast, along the coast it is formed of limestone of coral origin (Mjemah et al. 2009). The semi-confined aquifer has an average thickness of 100 m and consists of medium to coarse sand and sometimes gravel lenses, contained in a primarily clay matrix, and confined by a clay aquitard with a variable thickness of between 10 and 50 m (Mjemah et al. 2009). Approaching the coast, the aquitard is fragmented and the two aquifer formations are more hydraulically interconnected as a result. Quaternary aquifers are supported by a thick clay acquiclude that confines the underlying Neogene multi-layer aquifer at depths that can reach 1000 m (Mjemah 2007).

Analysis of saltwater intrusion is focused on the Quaternary aquifer formations, as they are most frequently used by the population. Because of the fragmentation of the clay aquitard in the range next to the coastline, the unconfined and semi-confined aquifers (Quaternary) are considered as a single aquifer.

4.3.3 Methods and Results

This first analysis consisted in evaluating the indicators IF1, IF2, and IF3, considering three different time intervals (2000–2002, 2005, and 2008–2010) and dividing the study area in two strips parallel to the coastline (inner strip and outer strip) with a width of 1.5 km. Time intervals were chosen according to the availability of data for each year and in a manner that ensured a homogeneous distribution of water points in the study area. The time interval 2000–2002 was considered in the analysis as the starting point because during that period the aquifer was characterized by relatively undisturbed natural conditions, groundwater exploitation in Dar es Salaam having begun just a few years before.

The data analyzed were obtained from 165 boreholes located within or just outside the study area (Fig. 4.5), and include Static Water Level (SWL), physical parameters of groundwater (T, EC, TDS) and chemical parameters of groundwater (Cl^-, Ca^{2+}, Mg^{2+}, Na^+, K^+). The historical data, collected from documents and reports prepared by public authorities (Ministry of Water and Irrigation—Drilling

Fig. 4.5 Borehole locations

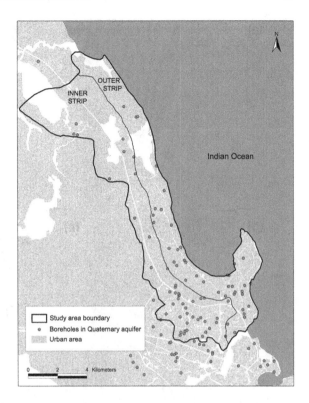

and Dam Construction Agency), researchers (Mato 2002; Mjemah 2007) and international cooperation agencies (JICA 2005), were integrated through a hydrogeological monitoring campaign (January–April 2010), using a multi-parametric probe, performed on a subgroup of 21 boreholes uniformly distributed in the study area.

When evaluating the indicator IF1 (average values of EC in the strip) only the boreholes with marine contamination were considered. High electric conductivity values for groundwater may be related to a state of saline contamination caused by various sources (i.e. anthropogenic pollutant leakage, soil leaching, and groundwater dissolution of evaporite rocks), and not necessarily attributable to seawater intrusion (Custodio and Bruggeman 1987; FAO 1997). Chemical analysis of groundwater ion correlation therefore allowed for the identification of a subset of 41 boreholes for which high EC values, and consequently high salinity levels, are due exclusively to marine contamination (Appelo and Postma 2005; FAO 1997; Bear et al. 1999). Values of EC for groundwater were divided according to FAO classifications (1992), which set the acceptable range of electric conductivity for irrigation water (freshwater: 0–2,000 μS/cm; brackish: 2,000–10,000 μS/cm; saltwater: 10,000–25,000 μS/cm; seawater: 25,000–45,000 μS/cm; brine: >45,000 μS/cm).

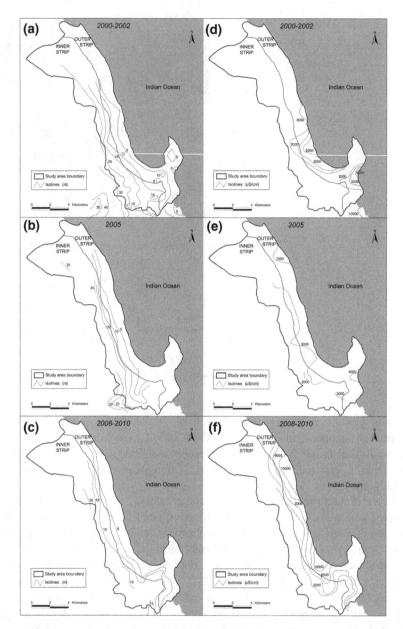

Fig. 4.6 Isophreatic contour lines (**a, b, c**) and Isoconductivity contour lines (**d, e, f,**) for the time intervals considered

According to this classification, water with EC values above 2,000 µS/cm is not considered potable and is not suitable for most crops. Indicator IF2 (maximum piezometric charge in the strip) was evaluated through reconstruction of the

Table 4.1 Indicator values

Time intervals	Indicators					
	IF1 (μS/cm)		IF2 (m)		IF3 (m)	
	Outer strip	Inner strip	Outer strip	Inner strip	Outer strip	Inner strip
2000–2002	3,091	1,604	15.00	30.30	1,500	2,380
2005	3,379	2,185	13.90	29.50	1,500	2,720
2008–2010	5,024	2,897	12.00	23.84	1,500	3,290

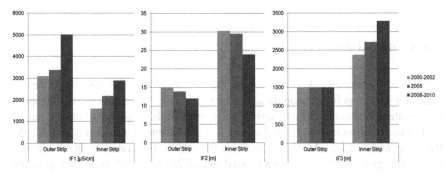

Fig. 4.7 Indicator trend in inner and outer strips

isophreatic contour lines, interpolating the point values of SWL (Fig. 4.6a, b, c). Similarly, the evaluation of the indicator IF3 (maximum distance from the coast of the isoconductivity contour line 2,000 μS/cm in the strip) was performed through the reconstruction of the isoconductivity contour lines based on the available EC values (Fig. 4.6d, e, f). Table 4.1 and Fig. 4.7 show the values of the indicators in the various time intervals and strips.

Although the available data have allowed for the analysis of a 10-year time period, primary results show a general increase in marine contamination over a large part of the study area. Increases in indicator IF1 seem to be a common trend in the outer strip, though it also occurs in the inner strip, where the mean EC value shifts from freshwater up to the brackish. Meanwhile, in the outer strip the mean EC value already corresponded to that typical brackish water in 2000, and presents an acceleration in the rate of increase between 2005 and 2008–2010. As regards indicator IF2, within the study area the water table varies from 0 to 30 m above mean sea level. Although this indicator is strongly dependent on the available point values of SWL, whose location may vary in different time intervals, analysis reveals a general decrease in the maximum piezometric level in both strips: almost 6.5 m in the inner strip and 3 m in the outer strip. This decrease is also observable by comparing the trend of the 5–10 m isophreatic contour lines in the different time intervals, which show a strong tendency to recede inland from 2000 to 2010.

Finally, as regards the indicator IF3, the trend of the isoconductivity contour lines shows the same tendency to recede inland. In fact, in the inner strip the mean distance of the isoconductivity contour line 2,000 µS/cm from the coastline increases by about 910 m from 2000 to 2010. Moreover, in the outer strip, where the indicator maintains the highest value (maximum width of the strip) in different time intervals, the trend of the isoconductivity contour lines shows an increase in EC values along the coastline, presenting in 2008–2010 areas with EC values greater than 10,000 µS/cm, corresponding to saltwater according to FAO classification (1992).

4.4 Conclusions

The survey methodology presented in this chapter seeks to identify different exposure scenarios for the population of Dar es Salaam's coastal area with respect to seawater intrusion in coastal aquifers as a result of the effects of climate change.

The use of historical data sets (Drilling and Dam Construction Agency Borehole Reports; Mato 2002; Mjemah 2007; JICA 2005) integrated with a specific hydrogeological survey campaign carried out in 2010, has allowed for a primary assessment of the identified indicators in a sample area during three time periods spanning the decade 2000–2010. Overall, the three selected indicators show a trend consistent with the evolution of seawater intrusion in the coastal area of Dar es Salaam identified by other authors. With reference to the area close to the shoreline (outer strip), the indicators clearly demonstrate progressive groundwater salinization and its possible relationship with decreasing piezometric levels caused by withdrawals and decreases in direct aquifer recharge. For the inner strip, the histogram analysis (Fig. 4.7) shows that the groundwater there has also been subject to salinization, even if only marginally (*brackish water* values), and evolution trends are much lower than in the outer strip. It was not possible to assess the effectiveness of the climate change variables (VCCs) or the possibility of identifying simple correlations with the IFs.

Critical analysis of the proposed methodology will only be possible after the assessment of many other parameters for the time periods considered, such as direct aquifer recharge due to rainfall infiltration. In order to do so, a broad collection of climatic data will be necessary, together with an evaluation of the temporal evolution of land cover (see Chap. 5). The IFs successfully tracked the evolution of seawater intrusion in the sample area over the 2000–2010 study period, and therefore their use may be extended to the entire project.

References

Adepelumi AA (2008) Delineation of saltwater intrusion into freshwater aquifer of Lekki Peninsula, Nigeria. Paper presented at the 3rd international conference on water resources and arid environments, Riyadh, 16–19 November 2008

Appelo CAJ, Postma D (2005) Geochemistry, groundwater and pollution. Taylor and Francis Group, London

Appleton B (2003) Climate changes the water rules. Dialogue on water and climate. Printfine Ltd, Liverpool

Araguás L, Custodio E, Manzano M (2004) Groundwater and saline intrusion: selected papers from the 18th Salt Water Intrusion Meeting. 18th SWIM, Cartagena, 31 May–3 June 2004

Arnell NW, Livermore MJL, Kovats S et al (2004) Climate and socio-economic scenarios for global-scale climate change impacts assessments: characterizing the SRES storylines. Glob Environ Change 14:3–20

Ataie-Ashtiani B, Volker RE, Lockington DA (1999) Tidal effects on sea-water intrusion in unconfined aquifers. J Hydrol 216(1–2):17–31

Barnett TP, Malone R, Pennell W et al (2004) The effects of climate change on water resources in the West: introduction and overview. Climatic Change 62:1–11

Bear J, Cheng AH-D, Sorek S et al (1999) Seawater intrusion in coastal aquifers—concepts, methods and practices. Kluwer Academic Publishers, Dordrecht

Bobba AG, Singh VP, Berndtsson R et al (2000) Numerical simulation of saltwater intrusion into laccadive Island aquifers due to climate change. J Geol Soc India 55(6):589–612

Brouyère S, Carabin G, Dassargues A (2004) Climate change impacts on groundwater resources: Modelled deficits in a chalky aquifer, Geer basin, Belgium. Hydrogeol J 12:123–134

Burns WCG (2002) Pacific Island developing country water resources and climate change. In: Gleick PH, Burns WCG, Chalecki EL, Cohen M (eds) World's water 2002–2003: the biennial report on freshwater resources. Island Press, Washington, pp 113–131

Clark GE, Moser SC, Ratick SJ et al (1998) Assessing the vulnerability of coastal communities to extreme storms: the case of Revere, Ma., USA. Mitig Adapt Strat Glob Change 3:59–82

Custodio E, Bruggeman GA (1987) Groundwater problems in coastal areas. Studies and reports in hydrology No. 35. UNESCO Press, Paris

Downing TE, Patwardhan A (2004) Vulnerability assessment for climate adaptation. In: Lim B, Spanger-Siegfried E (eds) Adaptation policy frameworks for climate change: developing strategies, policies and measures. UNDP, Cambridge University Press, Cambridge, pp 67–89

Eckhardt K, Ulbrich U (2003) Potential impacts of climate change on groundwater recharge and streamflow in a central European low mountain range. J Hydrol 284(1–4):244–252

FAO (1992) The use of saline waters for crop production. FAO irrigation and drainage paper 48. FAO, Natural Resources Management and Environmental Department, Rome

FAO (1997) Seawater intrusion in coastal aquifers: guidelines for study, monitoring and control. Water reports 11. FAO, Land and Water Development Division, Rome

Faye SC, Faye S, Wohnlich S et al (2004) An assessment of the risk associated with urban development in the thiaroye area (Senegal). Environ Geol 45:312–322

Füssel H-M (2007) Vulnerability: a generally applicable conceptual framework for climate change research. Glob Environ Change 17:155–167

Füssel H-M, Klein RJT (2006) Climate change vulnerability assessments: an evolution of conceptual thinking. Climatic Change 75:301–329

IPCC (2001) Climate Change 2001: impacts, adaptation and vulnerability. Contribution of Working Group II to the fourth assessment report of the intergovernmental panel on climate change. Cambridge University Press, Cambridge

IPCC (2007) Climate Change 2007: Impacts, adaptation and vulnerability. Contribution of Working Group II to the fourth assessment report of the intergovernmental panel on climate change. Cambridge University Press, Cambridge

IPCC (2008) Climate Change and Water. IPCC Technical Paper VI. http://www.ipcc.ch/pdf/
    technical-papers/climate-change-water-en.pdf. Accessed 12 Dec 2011
JICA (2005) The study on water supply improvement in coast region and dar es salaam peri-
    urban in the United Republic of Tanzania. Final Report. Japan International Cooperation
    Agency, Dar es Salaam
JICA (2012) The study on water resources management and development in Wami/Ruvu Basin in
    the United Republic of Tanzania. Progress Report (2). Japan International Cooperation
    Agency, Dar es Salaam
Kent PE, Hunt JA, Johnstone MA (1971) The geology and geophysics of coastal Tanzania.
    Geophysical paper 6. Natural Environmental Research Council, Institute of Geological
    Sciences, HMSO, London
Kjellen M (2006) From public pipes to private hands. Water access and distribution in Dar es
    Salaam, Tanzania. Intellecta DocuSys AB, Solna
Kortatsi BK (2006) Hydrochemical characterization of groundwater in the Accra plains of Ghana.
    Environ Geol J 50:299–311
Kortatsi BK, Jørgensen NO (2001) The origin of high salinity waters in the accra plains
    groundwaters. Paper presented at the 1st international conference on saltwater intrusion and
    coastal aquifers—monitoring, modeling, and management, Essaouira, 23–25 April 2001
Kundzewicz ZW, Mata LJ, Arnell NW et al (2008) The implications of projected climate change
    for freshwater resources and their management. Hydrol Sci J 53(1):3–10
Lu X (2006) Guidance on the development of regional climate scenarios for application in
    climate change vulnerability and adaptation assessments. National Communications Support
    Programme. UNDP-UNEP-GEF, New York
Margat J (2006) Les eaux souterraines: une richesse mondiale. BRGM Éditions/UNESCO, Paris
Mato RRAM (2002) Groundwater pollution in urban Dar es Salaam, Tanzania: assessing
    vulnerability and protection priorities. Dissertation, Eindhoven University of Technology
Mjemah IC (2007) Hydrogeological and Hydrogeochemical Investigation of a Coastal Aquifer in
    Dar es Salaam, Tanzania. Dissertation, Ghent University
Mjemah IC, Van Camp M, Walraevens K (2009) Groundwater exploitation and hydraulic
    parameter estimation for a quaternary aquifer in Dar es Salaam, Tanzania. J Afr Earth Sc
    55:134–146
Mtoni Y, Mjemah IC, Msindai K et al (2012) Saltwater intrusion in the quaternary aquifers of Dar
    es Salaam Region, Tanzania. Geol Belg 15:16–25
Oteri AU, Atolagbe FP (2003) Saltwater intrusion into Coastal Aquifers in Nigeria. Paper
    presented at the 2nd international conference on saltwater Intrusion and Coastal Aquifers—
    Monitoring, Modelling, and Management, Merida, Yucatan, 30 March–2 April 2003
Ranjan P, Kazama S, Sawamoto M (2006) Effects of climate change on coastal fresh groundwater
    resources. Glob Environ Change 80:25–35
Re V, Faye SC, Faye A et al (2011) Water quality decline in coastal aquifers under anthropic
    pressure: the case of a suburban area of Dakar (Senegal). Environ Monit Assess 172:605–622
Scibek J, Allen DM (2006) Modeled impacts of predicted climate change on recharge and
    groundwater levels. Water Resour Res 42. doi:10.1029/2005WR004742
Sherif MM, Singh VP (1999) Effect of climate change on sea water intrusion in coastal aquifers.
    Hydrol Process 13:1277–1287
Steyl G, Dennis I (2010) Review of coastal-area aquifers in Africa. Hydrogeol J 18:217–225
UN-HABITAT (2009) Tanzania: Dar es Salaam City Profile. UNION, Publishing Services
    Section, Nairobi
URT (2011) Dar e Salaam City environment outlook 2011. Division of Environment, Vice-
    President's Office, Dar es Salaam
Ziervogel G, Downing TE, Patwardhan A (2003) Linking global and local scenarios under
    climate change. SEI Poverty and vulnerability programme adaptation research workshop
    briefing paper. Stockholm Environment Institute, Stockholm

Chapter 5
Urban Sprawl as a Factor of Vulnerability to Climate Change: Monitoring Land Cover Change in Dar es Salaam

Luca Congedo and Michele Munafò

Abstract Urban sprawl is a major cause of environmental change, indirectly affecting climate processes on both the global and local scale and impacting the livelihoods of people who are directly dependent on ecosystem services. In the case of rapidly sprawling cities, land cover monitoring is a spatial planning requirement that must keep pace with urban growth, for the purpose of providing timely responses to environmental change and thus reducing people's vulnerability. Due to the lack of financial resources, Least Developed Countries need affordable methodology for rapid and effective land cover monitoring, suitable for low cost equipment. This chapter presents a methodology for monitoring land cover changes in Dar es Salaam, Tanzania, developed in the context of a project for the enhancement of local authorities' capacity to assess vulnerability to climate change and mainstream adaptation objectives into urban development plans. This methodology relies on the classification of free Landsat images and is implementable using open-source software, with the specific purpose of making sustainable the continuous assessment of urban sprawl for Dar City Council's planning services. The methodology phases are described, from preprocessing to processing. This includes the use of a free open-source plugin for QGIS, developed during the project, which allows for the semi-automatic classification of images. Classification results demonstrate the conspicuous urban growth of Dar es Salaam from 2002 to 2011, and provide insight into the relationship between urban sprawl and population growth.

Keywords Land cover change · Remote sensing · Landsat imagery · Semi-automatic classification · Dar es Salaam

L. Congedo (✉)
Department of Civil, Building, and Environmental Engineering, Sapienza University of Rome, Via Eudossiana 18, 00184 Rome, Italy
e-mail: ing.congedo.luca@gmail.com

M. Munafò
Italian Institute for Environmental Protection and Research, Via Brancati 48, 00144 Rome, Italy
e-mail: michele.munafo@isprambiente.it

S. Macchi and M. Tiepolo (eds.), *Climate Change Vulnerability in Southern African Cities*, Springer Climate, DOI: 10.1007/978-3-319-00672-7_5, © Springer International Publishing Switzerland 2014

5.1 Introduction to the Relationship Between Land Cover and Climate Change

The IPCC (2001) has defined land cover change as a non-climatic driver of environmental change because it influences climate change in several ways:

- soil has a major role in carbon fluxes and greenhouse gas emissions;
- land surfaces indirectly affect climate processes, because of the characteristics of materials on the ground;
- land cover change can alter the vulnerability of ecosystems to climate change.

More recently, the IPCC (2012) argued the need to include non-climatic factors in vulnerability assessments, and consequently the major role of spatial and land use planning was highlighted in terms of reducing exposure and vulnerability to environmental change. In this context, monitoring land cover change is fundamental to the development of effective policies for sustainability and adaptation to climate change, especially in the peri-urban fringe (Cardona et al. 2012).

The transition zone between urban and rural (i.e. the peri-urban interface) is often the location of the most rapid changes to the land cover of cities, driven by several causes including land tenure, migration, and other socio-economic motivations (Simon et al. 2006). This kind of unplanned development in the urban fringe, characterized by low-density and a mix of land uses, is generally defined as urban sprawl (EEA 2006).

Dar es Salaam is a rapidly developing city, and urban sprawl, which is related to high population growth rates but also to land speculation, represents a vast portion of the city's physical expansion (Kombe 2005).

The purpose of this study is to develop a methodology for land cover monitoring that allows Dar es Salaam planning services to assess urban changes with little effort and frequently enough to keep pace with rapid urban growth.

5.1.1 Land Cover Monitoring

Remote sensing and Geographic Information Systems (GIS) are very useful tools for assessing and monitoring the built-up expansion of the peri-urban interface and for mapping land cover (Brook and Davila 2000). Indeed, remote sensing can acquire data over large areas and detect sprawl patterns, as well as update spatial information over time (Chang 2012).

There are several approaches to land cover monitoring, including photo interpretation, field survey, and statistical, automatic, and semi-automatic classifications of remote sensing data. Several studies have used different methodologies of land cover classification to detect impervious surfaces, such as through the use of multi-spectral and hyperspectral sensors (Weng et al. 2008), medium spatial resolution (Fan et al. 2007), and high spatial resolution images (Myint et al. 2011). However, the choice of methodology for detecting land cover change depends on the spatial resolution and the temporal resolution of sensors, and will entail a varying degree of effort in terms of cost, time, output resolution, and accuracy (Richards and Jia 2006).

Dar es Salaam's local authorities are subject to several constraints with respect to technical equipment and financial resources (Kombe and Kreibich 2000). As emerged from interviews with local officers, the computers currently in use are old or low-spec, and the discontinuity of the electricity supply limits the use of electronic devices and the Internet. Also, only a few officers have GIS skills that are suitable for land cover monitoring and spatial analyses.

Therefore, in order to be tailored to the needs and resources of Dar City Council's planning services, the methodology for land cover monitoring developed within this study had to be:

- economical, in order to guarantee practicability for the municipalities;
- simple, in order to allow for regular updating of land use information;
- quick to execute, in order to detect the land cover changes in a very dynamic area.

The main objective of the developed methodology is to monitor land cover change, in particular at the peri-urban interface of Dar es Salaam, in order to identify the dynamics of urban sprawl.

5.2 Methodology for Monitoring Land Cover Change

In this study, a semi-automatic classification method was defined based on the subdivision of pixels from remote sensing images according to the spectral properties of the materials on the ground (the so-called spectral signature, i.e. the reflectance values at various wave lengths) (Richards and Jia 2006).

The present research concentrated on developing a methodology that satisfies the following requirements:

- based on, free or very low cost remote sensing images;
- image resolutions suitable for monitoring built-up and urban sprawl patterns;
- availability of images for past years, in order to monitor land cover change;
- semi-automatic classification, to reduce time and cost of land cover mapping;
- preprocessing and processing steps achievable with open-source software, to avoid the cost of proprietary software.

The workflow of the methodology is shown in Fig. 5.1. Landsat images, provided for free by the United States Geological Survey (USGS), were chosen because their multispectral resolution allows for the semi-automatic classification

Fig. 5.1 Workflow of the
developed methodology, for
preprocessing and processing
phases

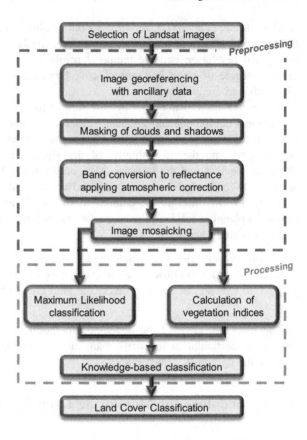

of impervious surfaces (Fan et al. 2007), and a large image archive for the past few
decades is available. Semi-automatic classifications allow for rapid and affordable
land cover monitoring.

5.2.1 Materials

In order to choose the most suitable satellite image for this study, several alternatives were considered. At present, there are many satellites with different resolutions, but most of them are commercial and unaffordable.

Multi-spectral and hyperspectral sensors like ALI and Hyperion, provided for
free by the USGS, can be used for land cover classification (Weng et al. 2008).
Unfortunately, no such images have been acquired over Dar es Salaam.

SPOT satellites are commercial, although some images are provided for free by
the European Space Agency (ESA) for research purposes. SPOT images can be
used for land cover classification (Huang et al. 2009). However, because they have

low spectral resolution, and their availability for Dar es Salaam over the past few years is limited, SPOT images are not the most suitable for our purposes.

Landsat images are provided for free by the USGS, in the context of a large image archive.[1] Therefore, Landsat was chosen for the land cover classification of the past few years.

Dozens of available images were acquired by Landsat 5 (TM) and 7 (ETM+) satellites, launched by the National Aeronautics and Space Administration (NASA) of the United States of America in 1984 and 1999 respectively. These two satellites have similar sensor characteristics, every image has seven multispectral bands with a 30 m spatial resolution, and a size on the ground of 170 km north–south by 183 km east–west (NASA 2011).

The developed methodology uses only the reflected solar energy bands, because they allow for the identification of materials according to the spectral response thereof (the thermal and panchromatic bands were excluded from the classification). Unfortunately, Landsat 5 acquisition was discontinued on 18 November 2011, due to electronic problems.[2] In addition, Landsat 7 images acquired after 2003 are affected by a technical problem with the Scan Line Corrector (SLC-off) causing gaps along the image, with stripes of null data.[3] Landsat 8, the successor of Landsat 7, was launched in February 2013 and is acquiring new images.[4]

5.2.2 Preprocessing

Preprocessing is the set of preliminary operations needed to eliminate or reduce radiometric and geometric errors (Richards and Jia 2006), and to prepare data for the classification process.

5.2.2.1 Georeferencing Images

Most of the images acquired had already been georeferenced by USGS, with a Standard Terrain Correction (Level 1T) that provides good geometric accuracy (less than 30 m). However, additional georeferencing was required for images with high cloud cover; the road network shape file and other images were used as references.

---

[1] Website http://earthexplorer.usgs.gov/ (accessed 14 June 2013) provides images acquired over Dar es Salaam since 1984. Landsat TM and ETM+ sensor characteristics are available in NASA (2011), p. 64.

[2] See http://www.usgs.gov/newsroom/article.asp?ID=3040 accessed 14 June 2013.

[3] Most images acquired over Dar es Salaam are affected by a high cloud cover percentage (the percentage of cloud cover over Dar es Salaam frequently above 50%). Twenty-three images with the lowest cloud cover were therefore selected. However, cloud-masking and mosaicing steps were included in the preprocessing workflow in order to produce a cloud-free image of the whole area.

[4] See https://lta.cr.usgs.gov/L8 accessed 14 June 2013.

5.2.2.2 Cloud Cover and Cloud Shadow Mask

A semi-automatic approach was used for masking clouds and their shadows. Landsat bands have a Digital Number (DN) range from 1 to 255. The developed method defines two thresholds for the DN of band 1 (blue) and band 6 (thermal) (Martinuzzi et al. 2007), based on the following steps:

1. In band 1, clouds are bright (high DN), therefore the minimum DN (DN_m) is found visually in the image where clouds are thinner, and the following rule of pixel selection is set: $DN_m \leq DN \leq 255$.
2. In band 6, clouds are generally colder than other surfaces (low DN), therefore the maximum DN (DN_M) is found visually in the image where clouds are thinner, and the following rule of pixel selection is set: $1 \leq DN \leq DN_M$.
3. The actual cloud mask is created by the intersection between these two rules, and an additional 3-pixel buffer is set around the detected cloud patches, in order to mask the thinner edge of clouds.
4. The cloud shadow mask is created with a maximum likelihood classification (see Sect. 2.3), selecting light and dark shadows.

The mask produced is a raster, where DN = 0 identifies masked pixels, in this case cloud and shadows pixels, and DN = 1 identifies non-masked areas.

5.2.2.3 Reflectance Conversion and Atmospheric Correction

The conversion from DN to *reflectance* and the atmospheric correction of Landsat images is required before classification and change detection (Song et al. 2001). Landsat records reflected solar energy for bands 1–5 and 7 and emitted energy for band 6. Prior to media output, image pixels are scaled to byte values, and it is therefore possible to calculate the *spectral radiance* at the sensor's aperture (L_λ), measured in watts/(meter squared * ster * μm), described in NASA (2011) as:

$$L_\lambda = G_{\text{rescale}} * Q_{CAL} + B_{\text{rescale}} \tag{5.1}$$

G_{rescale} is the rescaled gain (described in the Level 1 product header or ancillary data record) in watts/(meter squared * ster * μm)/DN, given by:

$$G_{\text{rescale}} = (LMAX_\lambda - LMIN_\lambda) / (Q_{CALMAX} - Q_{CALMIN}) \tag{5.2}$$

where: Q_{CAL} is the quantized calibrated pixel value in DN; $LMIN_\lambda$ is the spectral radiance that is scaled to Q_{CALMIN} [watts/(meter squared * ster * μm)]; $LMAX_\lambda$ is the spectral radiance that is scaled to Q_{CALMAX} [watts/(meter squared * ster * μm)]; Q_{CALMIN} is the minimum quantized calibrated pixel value in DN (corresponding to $LMIN_\lambda$) and in this case is equal to 1; Q_{CALMAX} is the maximum quantized calibrated pixel value in DN (corresponding to $LMAX_\lambda$) and in this case it is equal to 255. $LMIN_\lambda$ and $LMAX_\lambda$ values for every band are reported NASA (2011, p. 118).

$B_{rescale}$ is the rescaled bias (the data product *offset* contained in the Level 1 product header or ancillary data record) in watts/(meter squared * ster * μm), given by:

$$B_{rescale} = LMAX_\lambda - (LMAX_\lambda - LMIN_\lambda) / (Q_{CALMAX} - Q_{CALMIN}) * Q_{CALMIN}$$
(5.3)

Therefore substituting Eqs. (5.2) and (5.3) in Eq. (5.1):

$$L_\lambda = [(LMAX_\lambda - LMIN_\lambda) / (Q_{CALMAX} - Q_{CALMIN})] * (Q_{CAL} - Q_{CALMIN})$$
$$+ LMIN_\lambda$$
(5.4)

In order to reduce the variability between different images, spectral radiance is converted, as calculated in Eq. (5.4), to *planetary reflectance* (ρ_p, unitless) using the following equation (NASA 2011): $\rho_p = (\pi * L_\lambda * d^2) / (ESUN_\lambda * cos\theta_s)$, where d is the Earth-Sun distance[5] on the day of image acquisition; $ESUN_\lambda$ is the *mean solar exo-atmospheric irradiance* (a list of values for every Landsat band can be found in NASA (2011, p. 119); θ_s is the *solar zenith angle* (in degrees), which is the complementary of the sun elevation angle reported in the metadata file of each Landsat image.

As described by Zhang et al. (2010), *land surface reflectance* (ρ) can be estimated by the following equation: $\rho = [\pi * (L_\lambda - L_p) * d^2] / (T_v * F_d)$, where L_p is the path radiance due to atmospheric scattering; T_v is the atmospheric transmittance in the viewing direction; and F_d is the irradiance received at the surface expressed by Zhang et al. (2010) as: $F_d = E_b + E_{down}$.

E_{down} is the downwelling diffuse irradiance, and E_b is the beam irradiance defined as: $E_b = ESUN_\lambda * cos\theta_s * T_z$, where T_z is the atmospheric transmittance in the illumination direction. As originally described by Moran et al. (1992), the equation for converting *at-satellite radiances* to *surface reflectance* and correcting for both solar and atmospheric effects is:

$$\rho = [\pi * (L_\lambda - L_p) * d^2] / [T_v * ((ESUN_\lambda * cos\theta_s * T_z) + E_{down})]$$
(5.5)

The Dark Object Subtraction (DOS) is an image-based atmospheric correction technique, which does not require any field measurement during image acquisition. It is based on the assumption that some pixels within the image are in complete shadow (i.e. the *Dark Object*, which is assumed to have a reflectance value of 0.01), and their *at-sensor radiance* is due to L_p (Chavez 1996). Therefore the *path radiance* is calculated by Sobrino et al. (2004) as follows:

$$L_p = L_{min} - L_{DO1\%}$$
(5.6)

---

[5] In astronomical units. A free spreadsheet file reporting d for every day of the year is provided by NASA at http://landsathandbook.gsfc.nasa.gov/excel_docs/d.xls accessed 14 June 2013.

L_{min} is the minimum radiance corresponding to the minimum DN (DN_{min}) for which the sum of the number of pixels with DN $\leq DN_{min}$, is equal to 0.01 % of the total pixels composing the image. Therefore, once the DN_{min} value is identified, it is expressed by:

$$L_{min} = G_{rescale} * DN_{min} + B_{rescale} \qquad (5.7)$$

$L_{DO1\%}$ is the radiance of a Dark Object and is calculated by:

$$L_{DO1\%} = 0.01 * [(ESUN_\lambda * cos\theta_s * T_z) + E_{down}] * T_v / (\pi * d^2) \qquad (5.8)$$

The *path radiance* is obtained by substituting Eqs. (5.7) and (5.8) in Eq. (5.6), resulting in the following equation:

$$L_p = G_{rescale} * DN_{min} + B_{rescale} - 0.01 * [(ESUN_\lambda * cos\theta_s * T_z) + E_{down}]$$
$$* T_v / (\pi * d^2) \qquad (5.9)$$

Several studies explored different DOS models (DOS1, DOS2, DOS3, and DOS4) for calculating the variables T_v, T_z, and E_{down} (Song et al. 2001). The present study used the DOS1 model (Chavez 1996) because it is simple and efficient. The DOS1 model assumes no atmospheric transmittance loss, and corrects for spectral band solar irradiance and solar zenith angle, resulting in: $T_v = 1; T_z = 1$; and $E_{down} = 0$. Substituting these values for T_v, T_z and E_{down} in Eq. (5.9) results in:

$$L_p = G_{rescale} * DN_{min} + B_{rescale} - 0.01 * ESUN_\lambda * cos\theta_s / (\pi * d^2) \qquad (5.10)$$

Therefore Eq. (5.5) results in:

$$\rho = [\pi * (L_\lambda - L_p) * d^2] / (ESUN_\lambda * cos\theta_s) \qquad (5.11)$$

where L_λ is defined by Eq. (5.4) and L_p is defined by Eq. (5.10). Equation (5.11) converts DN directly to land surface reflectance by applying the DOS1 atmospheric correction.

5.2.2.4 Image Mosaic

Cloud masked images were processed in a mosaic, which replaces image gaps with pixels at the same location from other images. The mosaic process requires a variable number of images, depending on the amount and location of image gaps, which should be acquired within the same time period.

Often, image availability did not allow for the mosaicing of images acquired in the same month or season, so a higher amount of spectral variability was induced in the mosaic to be classified. Indeed, images acquired in different months of the year are affected by fluctuations in reflectance values, especially vegetation surfaces subject to phenological changes. Therefore, radiometric normalization is

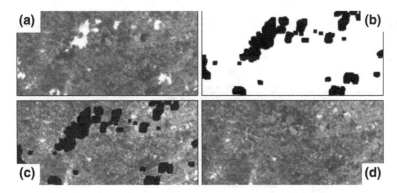

Fig. 5.2 Example of the mosaicing process steps: **a** the original cloudy image; **b** the creation of the cloud and shadow mask; **c** the application of the mask over the original image; **d** the image mosaic, with cloud and shadow gaps replaced

required in order to adapt the histogram of each image in the mosaic. Figure 5.2 shows an example of the image mosaicing process for two Landsat 5 images.

5.2.3 Classification Process

Land cover classification was performed through the supervised semi-automatic maximum likelihood algorithm, which assumes a multivariate normal distribution of the classes' probability (Richards and Jia 2006). Maximum likelihood has been employed in many land cover change studies (Song et al. 2001; Fan et al. 2007; Huang et al. 2009).

Using image processing software, a set of input training areas are collected in the image, representing the land cover classes; later, the software program calculates, for each class, the spectral signature statistics required by the maximum likelihood algorithm.

The maximum likelihood classifies each pixel of the image according to its the spectral similarity with the spectral signatures (Fig. 5.3).

5.2.3.1 Spectral Vegetation Indices

Vegetation indices are standardized methods based on band ratios that highlight vegetation dynamics (Song et al. 2001). In order to improve the land cover classification results, the following indices were calculated and integrated into a Knowledge-Based Classification:

- the *Normalized Difference Vegetation Index* (NDVI) is a combination of Red and Near Infrared wavelengths, where higher values correspond to better health of vegetation cover (Huang et al. 2009);

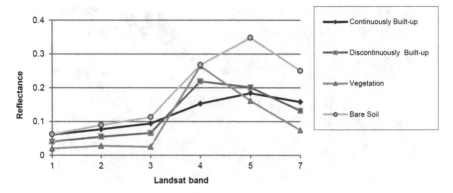

Fig. 5.3 Spectral signatures of some training areas collected in the image

- the *Enhanced Vegetation Index* (EVI) is also related to vegetation health, and seeks to correct for aerosol scattering, absorption, and background brightness through a combination of Blue, Red, and Near Infrared wavelengths (Soudani et al. 2006).

5.2.3.2 Knowledge-Based Classification

Knowledge-based classification is based on prior knowledge of the spectral, morphological or topological characteristics of land cover classes (Costa et al. 2010). In particular, the inputs for the knowledge-based classification were the maximum likelihood classification, NDVI, EVI, and Dar es Salaam boundary shapefile for limiting land cover classification to the administrative area.

Although the focus was on urban patterns, a total of six land cover classes were identified in the image, defined as follows:

- Continuously Built-up—a high-density urbanized class, identified directly by maximum likelihood classification;
- Discontinuously Built-up—a low-density urbanized class, characterized by a mix of urban and vegetation or soil pixels, identified directly by maximum likelihood classification;
- Full Vegetation—a vegetation class with NDVI values greater than or equal to 0.65;
- Mostly Vegetation—a vegetation class, identified by maximum likelihood classification or with NDVI values greater than 0.55 and lower than 0.65;
- Soil—identified directly by maximum likelihood classification;
- Water—identified directly by maximum likelihood classification or with EVI values lower than 0.

NDVI values vary according to the season, therefore these thresholds were adjusted according to the image acquisition date.

Table 5.1 Land cover classification results (in hectares)

	2002	2004	2007	2009	2011
Continuously built-up	8,415	10,025	10,447	12,370	14,808
Discontinuously built-up	8,098	9,134	12,509	17,318	23,678
Soil	102,079	95,732	76,011	57,385	66,791
Water	193	276	304	7	199
Full vegetation	14,887	13,172	14,905	26,751	18,195
Mostly vegetation	35,164	40,631	54,798	55,144	45,313

5.2.3.3 Semi-Automatic Classification Plugin for QGIS

Land cover monitoring through remote sensing requires specific software products, particularly for the semi-automatic classification of multispectral images, the methodology applied in this project. Most software products are commercial and their purchase costs might be unaffordable for the technical services of local authorities in LDCs. A few open-source (and free) programs could be used as an alternative—like Orfeo Toolbox and SAGA—but they lack some of the features that make classification easier. In order to close this gap, a Semi-Automatic Classification plugin (i.e. a program that adds features to software) has been developed for the open-source software QGIS.

This plugin (developed in Python) relies on other open-source software, and allows for the collection of training areas through a region-growing algorithm.[6] The collected training areas have the advantage of being spectrally homogeneous, and therefore allow for a better definition of land cover classes. The plugin allows for image classification using one of the following algorithms: Maximum Likelihood, Minimum Distance, and Spectral Angle Mapping.

The Semi-Automatic Classification Plugin can satisfactorily replace commercial software for land cover classification, making land cover monitoring more affordable.

5.3 Results of Land Cover Classifications

The land cover classification results for years 2002, 2004, 2007, 2009, and 2011 are listed in Table 5.1, and corresponding land cover maps are shown in Fig. 5.4.

As shown in Table 5.1, both the Continuously Built-up and Discontinuously Built-up classes have increased during the past few years, at a particularly fast rate from 2007 on. Major growth of the Discontinuously Built-up class, which is

---

[6] The Semi-Automatic Classification plugin uses the Orfeo Toolbox integration in QGIS (provided by the SEXTANTE plugin) in order to execute an image segmentation process, which allows for the selection of homogeneous pixels around a seed pixel of the image (region growing). This plugin is freely available through the QGIS interface, or at http://plugins.qgis.org/plugins/SemiAutomaticClassificationPlugin/ accessed 14 June 2013.

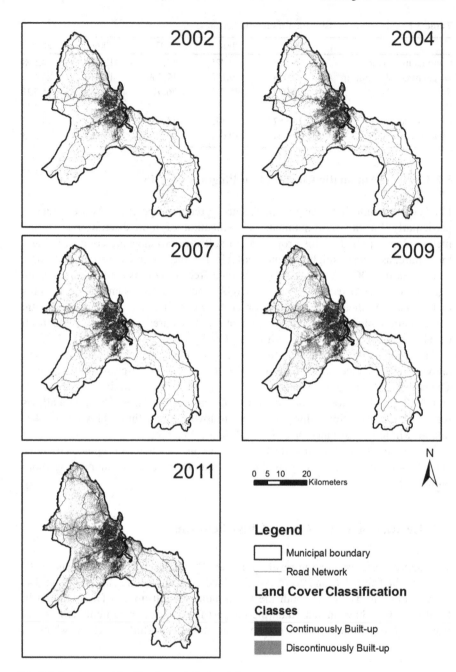

Fig. 5.4 Land cover classifications of Dar es Salaam

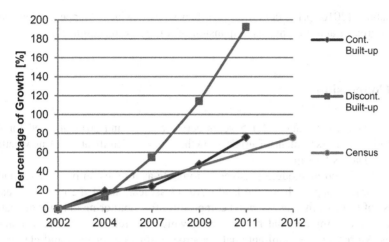

Fig. 5.5 Comparison of demographic and urban growth (2002-2012)

related to urban sprawl, has occurred along the main roads of Dar es Salaam. Over the years, parts of the low-density areas have transitioned to Continuously Built-up areas. The 30 m Landsat spatial resolution poses a challenge in urban sprawl monitoring, because a single image pixel could contain a mixture of cover types (Richards and Jia 2006), creating a mixed spectral signature (i.e. mixed pixel) depending on the composition and kinds of materials on the ground (Brook and Davila 2000). Moreover, the heterogeneity of materials on the ground complicates the definition of a spectrally distinct built-up class when small isolated patches of urban cover exist within a vegetated landscape, as is the case with sprawl (Shrestha and Conway 2011). Therefore, the Discontinuously Built-up area comprises both impervious and pervious surfaces.

During the classification process, the identification of training area pixels in Landsat images was difficult due to the rapid land cover change, and the lack of reference images with high spatial resolution. Land cover classification accuracy was calculated according to the methodology described in Congedo and Munafò (2012). For the 2011 land cover classification, the fuzzy overall accuracy is 72.0 %, and both the continuously and discontinuously built-up classes have high fuzzy accuracies for users (98.0 and 96.7 %, respectively) and producers (93.1 and 71.9 %, respectively).

Comparison between Demographic and Urban growth

In order to understand if urban and population growth are related, land cover classification results and census data were compared. Censuses were conducted in Dar es Salaam in 2002 and 2012, according to which the population grew from 2.5 to 4.4 million (United Republic of Tanzania 2013). Growth percentages since 2002 were calculated as a baseline (Fig. 5.5), with the following results: +76 % of Continuously Built-up; +192 % of Discontinuously Built-up; +75 % of Population.

It is worth noticing that demographic growth is lower than urban growth, especially for the Discontinuously Built-up class. This trend corresponds to the

UN-Habitat (2010, p.10) definition of urban sprawl, which "happens when population growth and the physical expansion of a city are misaligned".

5.4 Conclusions

Land cover maps produced for 5 years between 2002 and 2011 allowed for the monitoring and assessment of urban growth, and the identification of new settlements in Dar es Salaam.

The developed methodology is very affordable as it relies on free images and a semi-automatic approach for the land cover classifications of the past few years. The use of Landsat images is a good compromise for monitoring in terms of spatial accuracy and computational requirements, which are relevant for Dar's municipalities. Moreover, the semi-automatic approach allowed for the rapid classification of land cover, and rendered monitoring less time-consuming.

It is worth pointing out that the spatial resolution of Landsat images provides good results at the regional and municipal level. However, at the ward or local level, the spatial resolution is a constraint in the identification of land cover objects (a pixel is 900 m^2). Nevertheless, in accordance with the goals of this study (affordability, ease, and rapidity of execution), the use of free multispectral images was preferred over the use of high-resolution images (often not affordable, or with limited availability), as the latter does not allow for semi-automatic classification. Semi-automatic classification is crucial to reducing the time and effort required for photointerpretation and able to keep pace with the rapidity of urban development, thus ensuring the availability of updated land cover maps whenever they are needed by planning services.

The main challenge when using the developed methodology in Dar es Salaam is cloud cover, which necessitates mosaicing of multiple images. The mosaicing process increases the spectral variability of land cover classes, resulting in more classification errors, especially for classes with similar spectral signatures, such as Soil and Discontinuously Built-up. Moreover, logistical issues such as a weak internet connection can limit the download of large images, and limited hardware is a constraint for the creation of a municipal GIS.

The preprocessing and processing steps are achievable with open-source software, thus avoiding the cost of proprietary software; in particular, a plugin for QGIS was developed for the semi-automatic classification. This methodology is likely to be improved by the new Landsat 8, recently launched by NASA, whose images provide some additional multispectral bands useful for the detection of cirrus clouds.[7] Also, a valid alternative are Landsat are the ESA Sentinel-2 satellites, which have similar or better image characteristics.

---

[7] See http://landsat.usgs.gov/ldcm_vs_previous.php accessed 14 June 2013.

The land cover change results indicate the rapid expansion of built-up surfaces, particularly for the Discontinuously Built-up class along the main thoroughfares, as well as the gradual densification of central areas, from discontinuously to continuously built-up. This indicates urban sprawl, driven by considerable population growth (United Republic of Tanzania 2013), migration flows from upcountry, and land speculation (Kombe 2005).

Further studies (Congedo et al. 2013a, b, c) have highlighted the vicious circle between urban sprawl and climate change. The larger the sprawl area, the higher the consumption of environmental resources (e.g. more boreholes are drilled for water pumping), which in turn can worsen environmental conditions (e.g. water shortages), and push people to move to other places, fostering more urban sprawl.

References

Brook RM, Davila, J (eds) (2000) The Peri-Urban interface: a tale of two cities. School of Agricultural and Forest Sciences, University of Wales and Development Planning Unit, University College London, London

Cardona OD, van Aalst MK, Birkmann J, Fordham M, McGregor G, Perez R, Pulwarty RS, Schipper ELF, Sinh BT (2012) Determinants of risk: exposure and vulnerability. In: Managing the risks of extreme events and disasters to advance climate change adaptation. A special report of working groups I and II of the intergovernmental panel on climate change (IPCC). Cambridge University Press, Cambridge

Chang NB (ed) (2012) Environmental remote sensing and systems analysis. Taylor & Francis Group, Boca Raton

Chavez PS (1996) Image-based atmospheric corrections—revisited and improved. Photogramm Eng Remote Sens 62(9):1025–1036

Congedo L, Munafò M (2012) Development of a methodology for land cover classification validation. ACC DAR. http://www.planning4adaptation.eu/. Accessed 14 June 2013

Congedo L, Munafò M, Macchi S (2013a) Investigating the relationship between land cover and vulnerability to climate change in Dar es Salaam. ACC DAR. http://www. planning4adaptation.eu/. Accessed 14 June 2013

Congedo L, Munafò M, Macchi S (2013b) Land cover change and demographic growth: an estimation of population in Dar es Salaam using remote sensing. Paper presented at the international workshop bearing the brunt of environmental change: understanding climate adaptation and transformation challenges in African cities, Royal Holloway, University of London, London, 16–17 April 2013

Congedo L, Macchi S, Ricci L, Faldi G (2013c) Urban Sprawl e Adattamento al Cambiamento Climatico: il caso di Dar es Salaam. Paper presented at the XVI national conference of the Italian Society of Urban Planners, University Federico II of Naples, Naples, 9–10 May 2013

Costa GAOP, Feitosa RQ, Fonseca LMG, Oliveira DAB, Ferreira RS, Castejon EF (2010) Knowledge-based interpretation of remote sensing data with the interimage system: major characteristics and recent developments. In: The international archives of the photogrammetry, remote sensing and spatial information sciences ISPRS, vol XXXVIII-4/C7

EEA (2006) Urban sprawl in Europe—the ignored challenge. Report. EEA/OPOCE EEA/ OPOCE, Copenhagen

Fan F, Weng Q, Wang Y (2007) Land use and land cover change in Guangzhou, China, from 1998 to 2003, based on Landsat TM/ETM+ Imagery. Sensors 7:1323–1342

Huang S-L, Wang S-H, Budd WW (2009) Sprawl in Taipei's peri-urban zone: responses to spatial planning and implications for adapting global environmental change. Landsc Urb Plan 90(1–2):20–32

IPCC (2012) Managing the risks of extreme events and disasters to advance climate change adaptation. A special report of working groups I and II of the intergovernmental panel on climate change. Cambridge University Press, Cambridge

IPCC (2001) Climate change 2001: impacts, adaptation, and vulnerability. Contribution of working group II to the third assessment report of the IPCC. Cambridge University Press, Cambridge

Kombe W (2005) Land use dynamics in peri-urban areas and their implications on the urban growth and form: the case of Dar es Salaam. Tanzania Habitat Int 29(1):113–135

Kombe WJ, Kreibich V (2000) Reconciling informal and formal land management: an agenda for improving tenure security and urban governance in poor countries. Habitat Int 24(2):231–240

Martinuzzi S, Gould WA, Ramos OM (2007) Creating cloud-free Landsat ETM+ data sets in tropical landscapes: cloud and cloud-shadow removal. General Technical Report IITF-GTR-32. USDA Forest Service Pacific Northwest Research Station, Portland

Moran M, Jackson R, Slater P, Teillet P (1992) Evaluation of simplified procedures for retrieval of land surface reflectance factors from satellite sensor output. Remote Sens Environ 41(2–3):169–184

Myint SW, Gober P, Brazel A, Grossman-Clarke S, Weng Q (2011) Per-pixel vs. object-based classification of urban land cover extraction using high spatial resolution imagery. Remote Sens Environ 115:1145–1161

NASA (2011) Landsat 7 Science Data Users Handbook. Landsat Project Science Office at NASA's Goddard Space Flight Center in Greenbelt, Maryland. http://landsathandbook.gsfc.nasa.gov/pdfs/Landsat7_Handbook.pdf. Accessed 14 June 2013

Richards JA, Jia X (2006) Remote sensing digital image analysis: an introduction. Springer, Berlin

Shrestha N, Conway TM (2011) Delineating an exurban development footprint using SPOT imagery and ancillary data. Appl Geogr 31(1):171–180

Simon D, McGregor D, Thompson D (2006) Contemporary perspectives on the peri-urban zones of cities in developing areas. In: Simon D, Thompson D, McGregor D (eds) The Peri-urban interface: approaches to sustainable natural and human resource use. Earthscan, London

Sobrino J, Jiménez-Muñoz JC, Paolini L (2004) Land surface temperature retrieval from LANDSAT TM 5. Remote Sens Environ 90(4):434–440

Song C, Woodcock CE, Seto KC, Lenney MP, Macomber SA (2001) Classification and change detection using Landsat TM data when and how to correct atmospheric effects? Remote Sens Environ 75(2):230–244

Soudani K, François C, Maire GL, Dantec VL, Dufrêne E (2006) Comparative analysis of IKONOS, SPOT, and ETM + data for leaf area index estimation in temperate coniferous and deciduous forest stands. Remote Sens Environ 102(1–2):161–175

UN-Habitat (2010) State of the world's cities 2010/2011: bridging the urban divide. Earthscan, London

United Republic of Tanzania (2013) 2012 population and housing census: population distribution by administrative areas. National Bureau of Statistics, Ministry of Finance, Dar es Salaam

Weng Q, Hu X, Lu D (2008) Extracting impervious surfaces from medium spatial resolution multispectral and hyperspectral imagery: a comparison. Int J Remote Sens 29:3209–3232

Zhang Z, He G, Wang X (2010) A practical DOS model-based atmospheric correction algorithm. Int J Remote Sens 31:2837–2852

Chapter 6
Linking Adaptive Capacity and Peri-Urban Features: The Findings of a Household Survey in Dar es Salaam

Liana Ricci

Abstract Despite the increasing number of studies on urban vulnerability in African cities, there has been little research focusing specifically on the determinants of adaptive capacity. However, a greater emphasis is now being placed on the role that local responses and socioeconomic conditions play in determining vulnerability to climate stress or climate change. Such considerations have led to an increase in the attention paid to adaptive capacity, to the social context in general, and to the specific structural conditions that cause social and urban vulnerability. The purpose of this chapter is to provide knowledge regarding the identification of the determinants and attributes of adaptive capacity in the peri-urban areas of Sub-Saharan cities. The study describes a household survey conducted in Dar es Salaam with a focus on the natural resource systems upon which people depend. Its main contribution is the generation of knowledge on household characteristics and autonomous adaptation strategies, and a framework for better understanding the relationship between these two aspects. This relationship is the basis for understanding the nature and components of adaptive capacity in coastal Dar's peri-urban areas, and for identifying interventions that could be carried out by local institutions in order to support autonomous adaptation and environmental management practices and to improve adaptive capacity.

Keywords Peri-urban Africa · Livelihood strategies · Natural resource management · Autonomous adaptation · Dar es Salaam

L. Ricci (✉)
Department of Civil, Building and Environmental Engineering, Sapienza University of Rome, Via Eudossiana 18, 00184 Rome, Italy
e-mail: liana.ricci@uniroma1.it

S. Macchi and M. Tiepolo (eds.), *Climate Change Vulnerability in Southern African Cities*, Springer Climate, DOI: 10.1007/978-3-319-00672-7_6, © Springer International Publishing Switzerland 2014

6.1 Introduction

Rapid urbanization of Sub-Saharan cities is increasing human exposure to natural and anthropogenic hazards and is undermining the ability of communities and governments to cope with environmental stress and extreme weather events (Douglas and Alam 2006; Satterthwaite et al. 2007; Moser and Satterthwaite 2008; Simon 2010). Fast urban growth leads to the proliferation of unplanned settlements, which take the form of peri-urban informal areas that, in the last several decades, have accommodated most of the demographic expansion. Those processes have shaped highly fragmented and dynamic rural-urban interfaces, characterized by constantly changing land uses, activities, and social and institutional arrangements. Different definitions of peri-urban areas exist. Nevertheless, in the current debate there is growing recognition of the fact that rural and urban features coexist both within cities and beyond their limits (Allen 2001; Simon 2008; Adell 1999), and that dichotomous urban-rural systems are inadequate as regards dealing with processes of environmental and developmental change (Allen 2003). *Peri-urban* refers to the areas in cities where urban and rural features and processes meet, intertwine, and interact, and where there are mixed populations, important environmental services, and consumption of natural resources (Allen 2006).

The degree to which urban and peri-urban populations depend on natural resource-based activities plays a significant role in determining their vulnerability to environmental changes (Adger 1999) and poses institutional challenges for socio-ecological planning and vulnerability assessment (Eakin et al. 2010).

Vulnerability is a function of the exposure and sensitivity of a system to hazardous conditions and the ability, capacity, or resilience of the system to cope, adapt, or recover from the effects of those conditions (Smit and Wandel 2006). These definitions are represented by the equation:

$$\text{Vulnerability} = f(\text{Exposure, Sensitivity, Adaptive Capacity})$$

Interpretations of this function differ according to whether a greater emphasis is placed on social vulnerability or on exposure to environmental change hazards (Füssel and Klein 2006).

Some authors stress the uncertainties surrounding hazards (Adger and Vincent 2005), which are related to future changes and their impacts on people and the resources on which they depend, to the social and environmental systems that underpin vulnerability, and to the interaction between the two. Some of those authors argue that, in adaptation planning, the focus should be on building adaptive capacity, where adaptive capacity is defined as a

> system's ability to adjust to a disturbance, moderate potential damage, take advantage of opportunities, and cope with the consequences of a transformation that occurs (Gallopin 2006, p. 296).

For other authors this means

> the capacity to modify exposure to risks associated with climate change, absorb and recover from losses stemming from climate impacts, and exploit new opportunities that arise in the process of adaptation (Adger and Vincent 2005, p. 400).

According to these definitions, while exposure and sensitivity orient the potential impacts of climate change, adaptive capacity can be a major influence on its eventual effects. Therefore, adaptive capacity is an obvious focus for adaptation planning because it is the component of vulnerability most able to influence social systems that are coping with climate changes (Marshall et al. 2010).

While local people are often involved in planning processes and vulnerability assessments, the relevance of those processes as regards people's livelihood strategies and environmental management practices (the context of their vulnerability) is often neglected (Dodman and Mitlin 2011).

This chapter assumes that adaptive capacity is crucial, or rather that the focus should be on improving people's capacity to cope with the impacts of environmental stress (e.g. focusing on livelihoods and capabilities) rather than on specific interventions to prevent or reduce specific impacts (e.g. introduction of drought-resistant crops in response to decreasing water availability, infrastructure provisioning, etc.).

Autonomous adaptation (defined as people's spontaneous actions undertaken to cope with environmental changes) and planned adaptation (institutional programmes, policies, projects, and initiatives) are interlinked and

> embedded in governance processes [reflecting] the relationship between individuals, their capabilities and social capital, and the government (Adger and Vincent 2005).

The hypothesis of the present research is that local institutions (at the political and practical level) can increase adaptive capacity by supporting and orienting autonomous adaptation, modalities of accessing resources, and environmental management practices.

The main objective of the research is to identify the basic elements of autonomous adaptation in peri-urban areas in order to inform local institutional interventions that are tailored to the needs of peri-urban communities in Sub-Saharan Africa.

A pilot study conducted in peri-urban Dar es Salaam analyzed livelihoods, natural resource management, and autonomous adaptation strategies undertaken at the household level. It has emerged that *informal* environmental management and access to resources are, in many cases, the basis for autonomous adaptation and vulnerability reduction, as they facilitate diversification of livelihoods and modalities of accessing resources (water, land, etc.). These informal practices and strategies are mainly based on social relations and networks, which constitute *platforms of action* (Simone 2004). These platforms, which enable adaptation and sustain people's livelihoods, represent in some cases the extension or the substitution of infrastructure such as water pipelines, financial services, social services, and communication technologies. The functioning of these platforms of action depends on the characteristics of peri-urban households (e.g. income, location, livelihoods strategies). Those characteristics may constitute an opportunity or a limit for autonomous adaptation. On the other hand, autonomous adaptation practices might have negative or positive impacts on household characteristics. These two types of relationship are the basis upon which local institutions can intervene to build adaptation strategies and mainstream adaptation into urban planning and environmental management.

6.2 Selecting the Case Study: Dar es Salaam

Dar es Salaam was selected for its environmental problems and characteristics, the expansion of its peri-urban areas (the largest city in Tanzania and the 3rd fastest growing in Africa), its high projected growth rate (5.12 % between 2015 and 2020, according to World Population Prospects 2011 (UN 2010)) and the high percentage of informal settlements (70 %, according to Kombe (2005) and Lupala (2002), 80 % according to Kironde (2006)). According to data collected by UNDP (McSweeney and New 2008) and the IPCC (2007a), the main effects of climate change in Dar es Salaam are flooding, sea level rise, drought, and changes in rainfall patterns. Furthermore, global climate change scenarios indicate that the temperature will continue to rise and rainfall will increase in the years to come (IPCC 2007b). The downscaling of these scenarios to the local level has led to many uncertainties, but a few studies have highlighted evident trends of change at the local level, including salt water intrusion (Faldi 2011), decreasing rainfall (from about 1200 mm per year in the 1960s to about 1000 mm in 2009), and increasing temperatures, with the greatest temperature increase recorded in the decade 1999–2008 (URT 2011). In addition, urban development and changes in socio-economic conditions in peri-urban areas are likely to exacerbate the physical drivers of the changes mentioned above, impacting the probability of climate events and stresses and orienting the positive or negative impacts thereof.

6.2.1 Scale and Rationale of the Analysis

A survey was developed based the hypothesis that vulnerability in peri-urban areas depends on several factors, including type and magnitude of the local impacts of environmental changes, rural-urban dynamics, land-use patterns and urban fabric, local autonomous capacity to cope with environmental changes, and institutional capacity for environmental management and urban development planning. Of those factors, the questionnaire focused on autonomous adaptation and the related peri-urban household characteristics.

The household questionnaire was built on four main assumptions and aimed to investigate the key factors of adaptive capacity in peri-urban Dar es Salaam. The first assumption is that understanding the rural-urban interaction, the economic flows, the flow of resources, and socio-cultural relations that characterize peri-urban areas is fundamental when attempting to identify viable adaptation options. The rural-urban interface is also the place with the most potential for positive change, due to the many forces that come together in this space (e.g. multifunctional urban agriculture) (Erling 2007; Simon 2008). The dynamic and diversified rural-urban environment provides opportunities for change that could be useful for adaptation, such as livelihood diversification, access to services, reuse of waste in agriculture and waste recycling, and greater access to information and decision-making (Dàvila 2002; Allen 2003). Therefore, it was assumed that modalities of accessing resources (e.g. shorelines, the sea, raw materials) and environmental management are the basis upon which individuals choose autonomous adaptation options. These, in turn, might generate obstacles to or opportunities for adaptation, thus potentially undermining or enhancing adaptive capacity and having considerable effects on people's vulnerability. Fourthly, it was assumed, that the autonomous adaptation practices that are implemented and the observed environmental changes related thereto must also be taken into consideration when identifying priority adaptation actions (COP 7 Decision 28/CP.7).

On the basis of these assumptions, the questionnaire, which was designed and conducted in 2009, was divided into four thematic sections that were identified through a review of the literature on peri-urban areas and discussion with researchers from ARDHI University in Dar es Salaam. Those four areas are as follows:

• Rural-urban interactions;
• Modalities of access to resources (land, water, energy, etc.);
• Environmental management (water, waste, soil, etc.);
• Environmental change (observed environmental transformations and autonomous adaptation practices and strategies).

6.2.2 Sample Location and Questionnaire Administration

The survey was conducted on the northern border of Dar es Salaam. Forty household questionnaires were administered in four different wards (ten in each ward) located in the Kinondoni Municipality: Bunju, Kunduchi, Kawe, and Msasani (Fig. 6.1), where Msasani was an *urban* control case located very close to the core of the city. The wards were selected through a series of field visits, a review of the literature on Dar's peri-urban areas and discussion with academics from Ardhi University.

The sampled wards were selected according to several criteria, including the coexistence of urban and rural activities (agriculture, livestock, businesses, schools, and transport), the presence of informal settlements and activities, low-medium settlement density (each plot between 0.08 and 6 ha), and close proximity to major natural resources (rivers, ocean, wetlands, forests, etc.). In addition, sample areas were selected in a manner that guaranteed a mix of environmental characteristics

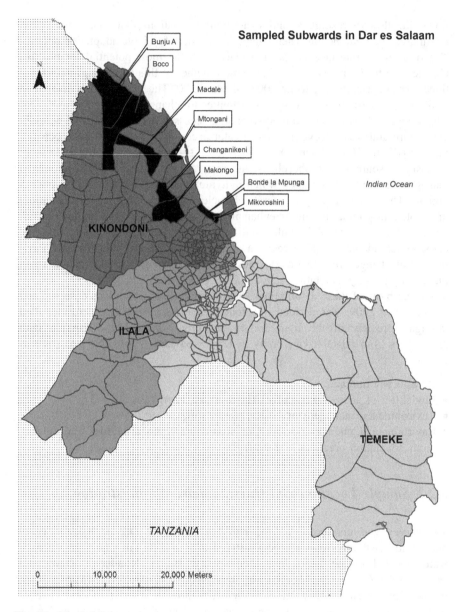

Fig. 6.1 Sampled Sub-wards

(coastal and inland areas with different morphologies). Sample households were then selected such that the sample reflected socio-economic and cultural hetero-geneity, while individual households were chosen if characterized by stability of settlement (having extensive knowledge of resources and local development dynamics) and dependence on both urban and rural activities and resources.

Key results of the data analysis are discussed below, in order to highlight the relationships between peri-urban characteristics and autonomous adaptation practices, which are the target of institutional interventions aimed at increasing adaptive capacity.

6.3 Results: Adaptation and Environmental Management in Dar es Salaam

While the limited number of questionnaires was not conducive to quantitative analysis, it did allow for observation of the behavior and characteristics of households in peri-urban areas. This data was then the basis for the development of a more comprehensive and targeted survey, performed in Dar between July and October 2011.

The data collected through the questionnaire facilitates an understanding of how autonomous adaptation and environmental management practices interact with the characteristics of peri-urban households and areas, and therefore how local institutions can intervene in these interactions to support autonomous adaptation and enhance adaptive capacity.

The data analysis is organized in two parts. In the first section, the characteristics of peri-urban households, their modalities of accessing various resources, and environmental management practices are described (Table 6.1). The second part then incorporates autonomous adaptation strategies (Table 6.2).

6.3.1 Peri-Urban Households' Characteristics

6.3.1.1 Rural-Urban Interaction

Responses concerning the rural-urban interaction suggest that the development of peri-urban areas is not caused exclusively by rural-to-urban migration, but also by urban-to-peri-urban migration. The latter was often induced by upgrading programmes undertaken in areas closer to the city center, from which several interviewees had migrated. In the three sampled peri-urban wards, the majority of households surveyed were originally from the same municipality, while only a few families were from other regions or municipalities.

There are different causes of migration. Almost half of the respondents indicated work or family as their reason for moving to a peri-urban area. Such areas attract people seeking a place where they can undertake or continue rural activities (e.g. the possibility of practicing agriculture or free husbandry) while also remaining close to urban dynamics, benefits, and facilities. Movements between peri-urban areas and the inner-city, and vice versa, occur frequently and imply flows of people,

Table 6.1 Peri-urban households' characteristics

Characteristic	Options
Livelihood strategies or income/non-income generating activities[a]	Agriculture and other rural activities as main livelihood activity
	Agriculture or other activities dependent on natural resources as complementary livelihood activity
	Urban employment only (not dependent on natural resources)
	Urban employment (not dependent on natural resources) and other secondary activities as main livelihood activity
	Urban and rural employment contribute almost equally to livelihood
Modalities of water access (options for water provisioning mainly for domestic use)	Street vendors or water tank truck only (A)
	Street vendors or water tank truck and water pipeline (A/I)
	Water pipeline only (I)
	Stream/river or other natural source only (A)
	Multiple source: combination of more than one modality (A) or (A/I)
Modalities of land access (options for right to land use or occupancy)	Title deed (leasehold or customary) (I)
	Without title deed (A)
	Indirect title deed (rent/purchase land from owner with title deed) (A/I)
	Indirect without title deed (rent/purchase land from owner without title deed) (A)
Modalities of waste management (solid waste and wastewater)	Public or private waste collection and management system (I) or (A)
	Autonomous waste collection and management system (burning, abandoning, burying, recycling, composting) (A)
	Combination of autonomous and organized waste collection and management systems (A/I)
Modalities of accessing other resources, services, and infrastructure	Electricity network (I)
	Electricity network combined with other sources (A/I)
	Other combined sources of energy (charcoal, kerosene, gas) (A)
	Private transport services (A/I)
Environmental location	Coastal (close to shoreline)
	Wetland
	Plateau
	Lagoon (mangroves)
	Close to a body of water
	Swamp

(continued)

Table 6.1 (continued)

Characteristic	Options
Urban location	Close to transportation hubs
	Close to main urban services (education, health, etc.)
	Close to main markets (Kariakoo, Mwenge, Tegeta, etc.)

[a] With regard to livelihood activities, the type autonomous/informal or institutional/formal has not been indicated, since for each type both are possible and simultaneously present

Table 6.2 Linking adaptive capacity and peri-urban features

Type of strategy	Adaptation practices
Change livelihood	Change crops
	Increase arable surface area
	Introduce additional economic activities
	Betterment or intensification of agriculture (e.g. change farming techniques, use fertilizer)
Changes modality of accessing resources (e.g. water)	Dig in humid areas or close to rivers to collect water
	Multiply sources for resource access
	Take out a loan
Change the environment (e.g. change soil morphology)	Build small embankments to preserve water in water bodies
	Dig channels for water drainage
	Make changes to house structure
	Build a new house
Move to another area	To areas with more space for agriculture
	To areas with better soil fertility
	To areas closer to main roads or transport hubs (for small shop/business)
	To region of origin
Change type of livelihood (e.g. from more to less dependent on natural resources)	Switch from agriculture to livestock
	Cease agriculture and/or livestock to undertake activity less dependent on natural resources (e.g. small business, trade)
	Switch from fishing to agriculture
	Search for temporary job

resources, information, commodities, production inputs, and decision-making power (Tacoli 1998). This means that a plurality of activities and physical patterns exist that are based on urban and peri-urban interdependencies. In some cases, movement to peri-urban areas is prompted by extreme climatic events (e.g. flooding) or by worsening environmental conditions (pollution, urban development, environmental change). The strong relationship between urban and peri-urban areas is also demonstrated by the large number of people who travel to Dar's city center on a weekly basis. Rural-urban linkages and dependence on natural resources are further evinced by people's primary livelihoods, which are often based on agriculture and livestock, and supplemented with other local informal activities (e.g. petty trade). This is also one of the reasons the majority of respondents wish to live in environments with *free* spaces, while only a few respondents wish to move to a more urbanized area with better infrastructure and facilities.

6.3.1.2 Access to Resources and Environmental Management

Modalities of resource access and of environmental management have a considerable effect on the vulnerability of peri-urban residents. Identification of resource use and the management regime is crucial to understanding households' capacities to interact with the environment, substitute or integrate facilities with their practices, and cope with the lack of water supply, waste management, and other services. Access to water, land, energy, shorelines, the sea, raw materials, and services constitutes in each case a determinant of autonomous adaptation, which is mainly an *informal* process based on social networks and direct relationships with the environment.

6.3.1.3 Land Titling and Land Formalization

In this framework, land tenure arises as an important issue linked to sources of income and food, and as a critical issue in the implementation of urban planning processes and actions. There is pressure in peri-urban areas, especially from city residents, to acquire and subdivide land for urban development (Kironde 2001, 2003b), even though agriculture is still an important source of income and food in those areas. Among respondents, nearly two thirds do not have land title, while the others have a title deed (leasehold), and only a few people claim to have customary title (right of occupancy derived from ancient communities). In addition to various categories of formal land tenure, informal tenure is also recognized as modality of accessing land (Kironde 2003a). However, though they are legally recognized, informal and customary tenure are becoming increasingly uncommon, and are not protected if other interests arise in urban development (Kironde 2004).

Informal tenure is the modality of land access on which most respondents rely. Although the majority of respondents do not hold a land title, they have usually purchased the land they use. While the linkages between vulnerability to environmental change and land tenure are complex and indirect, land acquisition processes certainly have a variety of implications for autonomous adaptation practices. Those processes orient land use changes that may undermine or facilitate autonomous adaptation practices and institutional adaptive interventions.

It is widely argued that security of tenure is mainly guaranteed by land formalization processes and land titling, which are also considered basic strategies for reducing vulnerability and increasing adaptive capacity. Yet questionnaire responses suggest that peri-urban residents, even if they have the economic resources for land survey and registration, opt mainly for informal tenure. The preference for informal tenure is also evinced by reactions to upgrading programmes (e.g. CIUP—Community Infrastructure Upgrading Program), where in many cases people decide to move to another area and take advantage of the opportunity to sell their plot, rather than remain in a formalized area. Even if land titling might confer benefits (access to credit, investment in property, access to infrastructure and services, etc.), the results of the present study indicate that increased security of tenure does not necessarily require formal tenure regularization. Informal legitimating processes seem to be more trustworthy than legal formalization because they allow flexibility in land supply and meet the demands of changing livelihoods. If adequate security of tenure provides incentives for good land and resource management and reduces vulnerability, flexibility also allows residents to cope with increasing land use changes, migration, the associated competition for land, and conflict resulting from environmental changes and intensified extreme events (Quan and Dyer 2008).

6.3.1.4 Accessing Water and Other Resources

Access to water, energy, and services such as waste management is also based on *informal* processes, social networks, and direct relationships with the environment. Among the sampled peri-urban areas, only Kunduchi ward has water supply infrastructure (water pipeline). Nevertheless, due to the unreliability of water services (lack of water, damaged pipelines, etc.) all households have several alternative modalities of accessing water. In fact, the main sources of water are street vendors or water tank trucks (2/3 of sampled households). Other households use natural sources such as streams, springs, ground pits, and wells. Access to water is highly diversified, and most households combine multiple access modalities, including natural sources, pipelines, neighbor's source, water sellers, and rainwater harvesting, among others (Fig. 6.2). This type of diversification also occurs with respect to water management and water storage systems, with the latter accomplished using small buckets (20 L) or underground concrete tanks. Even access to energy is widely diversified. Only a few of the sampled households have access to electricity, while most of them use charcoal for cooking and kerosene for lamps, as well as gas and other energy sources.

As regards environmental management, which also includes waste management, only a few households use an *organized* collection system, while the majority manage waste autonomously by burning, abandoning, and burying waste, or by recycling useful materials and composting organic waste to be reused as fertilizer (Figs. 6.3, 6.4). Others also collect materials such as plastics or metals for sale to companies or individuals that can treat or reuse them. Concerning sewage, no one uses a collection and disposal system; wastewater is usually abandoned in pits until saturation (pit latrine; rarely a septic tank is used). There are also

Fig. 6.2 Water sources

Fig. 6.3 Plastic collection
(reproduced from Ricci 2011)

informal groups and community-based organizations for maintenance of roads, canals, and other common spaces.

Table 6.1 summarizes the main features of peri-urban households identified through the household questionnaire analysis. These include livelihood activities, options for resource access and environmental management, the presence of infrastructure and facilities, and household location. Options undertaken autonomously (informally) are designated with (A), while the options provided by institutions are denoted (I), and those options dependent on both autonomous and institutional initiatives are identified as (A/I).

Fig. 6.4 Composting
(reproduced from Ricci 2011)

6.3.1.5 Autonomous Adaptation

Almost all respondents noted changes in water availability, soil fertility, soil aridity, air humidity, and rain patterns. Results show that water availability has declined significantly in recent years. Rivers that formerly flowed year-round have become seasonal, and the water previously drawn from shallow pits in the wetlands or near rivers has decreased, requiring deeper digging. Other significant changes were observed both in the amount of rain and in normal seasonal rainfall patterns. Furthermore, a significant erosion of the coastline (in some places as much as 100 meters in the last 30 years) has been observed and linked to changes in village morphology and fishing activities (in Kunduchi, the Mtongani fish market area). These and other environmental changes result from a complex set of factors that includes anthropogenic pressures and inadequate local environmental and urban policies as well as global environmental change. Different strategies for coping with environmental changes are being implemented (Figs. 6.5, 6.6). Because of the decrease in water availability, many people have changed crop systems (e.g. moving from rice to cassava, which requires less water) or have decided to stop farming and start breeding livestock. Most respondents, who have been observing rapid and significant environmental changes in recent years, are contemplating plans for coping with further deterioration of environmental conditions that go beyond immediate reactive solutions. They are considering strategies such as change of employment, and transition from subsistence activities dependent on natural resources to activities only partially or indirectly dependent on them (e.g. trade or small business). In some cases, respondents have even contemplated moving to another area or returning to their native rural region.

These strategies are being considered not only in response to exasperation of environmental problems, but also in the event of increased population and new urban developments, which would interfere with ordinary practices and activities. This response is partly linked to the causes to which respondents attribute

Fig. 6.5 Adaptation practice: river embankment (reproduced from Ricci 2011)

Fig. 6.6 Adaptation
practice: shallow pit for
harvesting water (reproduced
from Ricci 2011)

environmental change, including changes in land use, local anthropogenic actions
(mainly urban development), and global environmental change. In addition, a few
respondents attribute on-going changes to weak institutions and mismanagement
of resources. Table 6.2 shows the autonomous adaptation strategies divided into
five types, which are highly dependent on the above-mentioned household
characteristics.

6.3.2 A Framework for Assessing Adaptation Options

The foregoing analysis demonstrates that peri-urban households' adaptive capacities are linked to the possibility of diversifying sources of income and modalities of accessing water, land, and other resources. Furthermore, most modalities of resource access, environmental management, and adaptation practices are undertaken in a predominantly autonomous (informal) way, and depend on the presence of both urban and rural features. This rural-urban mix allows individuals to develop hybrid livelihood strategies in which rural-urban and formal-informal practices complement, subsidize, or support each other.

The primary contribution of this study is a better understanding of the relationship between the characteristics of peri-urban households and their autonomous adaptation strategies. By relating autonomous adaptation strategies to such characteristics, one can obtain four types of interaction between the two sets of information. First, autonomous adaptation practices impact the characteristic features of the peri-urban, either in a positive or negative way. Second, those features function as both opportunities and constraints for the diverse adaptation practices peri-urban dwellers can undertake to cope with changes in their living environments (Fig. 6.7).

The first two types of interaction evince how location, modalities of resource management and access, availability of services, and economic activities can inhibit or promote a particular adaptation practice. Specifically, a given characteristic generates an *opportunity* when it favors a particular autonomous adaptation practice (e.g. being located near large markets or major transportation routes facilitates commercial activities such as livelihood diversification being located near a river makes it easy to extract water from fairly shallow wells). *Obstacles*, on the other hand, arise when a given characteristic inhibits or impedes an adaptation practice (e.g. being located near a transportation hub or major market may prevent the expansion of on-site agricultural land as a result of higher population density).

The other two types of interaction demonstrate the kinds of effects that autonomous adaptation practices can have on the characteristics of peri-urban households (positive and negative impacts). In particular, a *positive impact* occurs when an adaptation practice improves environmental or economic conditions (e.g. using organic fertilizers in urban agriculture or increasing the cultivated acreage to cope with soil aridity or decreasing soil fertility could facilitate improved waste management through organic waste recycling, thus having a positive environmental impact). A *negative impact* occurs when the adaptation practice introduces risks or stresses for natural resources (e.g. the construction of embankments across drying river beds in order to preserve water and use it for agricultural and domestic purposes could damage the river ecosystem and exacerbate the effects of a changing climate on natural resources and people, by restricting access to water).

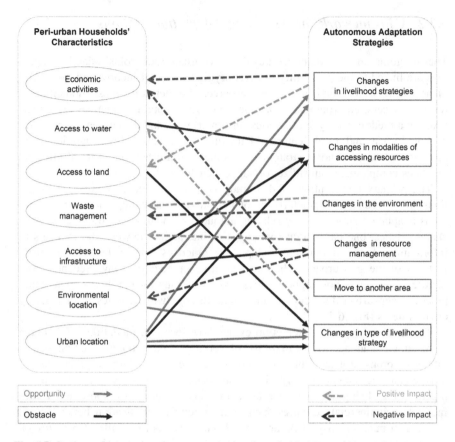

Fig. 6.7 Reciprocal interactions between peri-urban household characteristics and autonomous adaptation strategies (reproduced from Ricci 2011)

6.4 Conclusions

Investigation of the opportunities and obstacles arising from household characteristics can contribute to an understanding of how different modalities of resource access and environmental management, spatial location, facilities, and economic activities might promote or limit specific autonomous adaptation practices.

The analytical framework (Fig. 6.7) proposes a method of identifying the main factors on which adaptive capacity depends within peri-urban households. It emphasizes the interconnectedness of formal and informal access to resources and environmental management, and their importance for people's livelihoods and adaptation strategies.

This framework is particularly useful for institutions seeking to identify the crucial elements of adaptive capacity enhancement and to set priorities in the adaptation process. This is essential when considering adaptation options in concrete and specific social and environmental contexts and when operationalizing adaptation option assessments.

The case study illustrates that investigation of autonomous adaptation practices reveals a broad spectrum of options (explicit and implicit adaptation practices) from which institutions can draw valuable lessons in terms of identifying potential entry points for effective institutional adaptation options. Most adaptation practices are connected in a complex livelihood strategies framework, and different types of barriers to adaptation and livelihood strategies can be identified.

Nevertheless, the study has some limitations. It has yielded limited information on the role of local institutions in autonomous adaptation strategies. Also, the diverse modalities of resource access that have been identified require further investigation in order to provide valuable information on key factors of peri-urban adaptive capacity.

Indeed, further exploration of the applications of this framework would improve the analysis of the institutional dimensions of adaptation. It could also provide a basis for understanding adaptation conflicts between different actors, or could be used to analyze the costs associated with the development and implementation of adaptation policies and actions. Finally, the different positions of autonomous and planned adaptation in the livelihood strategies could be also investigated.

References

Adell G (1999) Theories and models of the peri-urban interface: a changing conceptual landscape. Literature review Strategic environmental planning and management for the peri-urban interface University College London, Development Planning Unit London: University College London

Adger WN (1999) Social vulnerability to climate change and extremes in coastal Vietnam. World Dev 27(2):249–269

Adger W, Vincent K (2005) Uncertainty in adaptive capacity (IPCC Special Issue on Describing Uncertainties in Climate Change to Support Analysis of Risk and Options). Comptes Rendus Geosci 337(4):399–410

Allen A (2001) Urban sustainability under threat. The industrial restructuring of the fishing industry in the city of Mar del Plata, Argentina. Dev Pract 11(2–3):152–173

Allen A (2003) Environmental planning and management of the peri-urban interface: perspectives on an emerging field. Environ Urban 15(1):135–148

Allen A (2006) Understanding environmental change in the contest of rural-urban interaction. In: Gregor M, Simon D, Thompson D (eds) The peri-urban interface approaches to sustainable natural and human resource use. Earthscan, London

Dàvila JD (2002) Rural-urban linkages: problems and opportunities. Espaço e Geografia 5(2):35–64

Dodman D, Mitlin D (2011) Challenges for community based adaptation: discovering the potential for transformation. J Int Dev 25:640–659. doi:10.1002/jid.1772

Douglas I, Alam K (2006) Climate change, urban flooding and the rights of the urban poor in Africa: key findings from six African cities. Action Aid, London

Eakin H, Lerner A, Murtinho F (2010) Adaptive capacity in evolving peri-urban spaces: responses to flood risk in the Upper Lerma River Valley, Mexico. Glob Environ Change 20:14–22

Erling G (2007) Multifunctional agriculture and the urban-rural interface. In: Laband DN (ed) Emerging issues along urban-rural interfaces II: linking land-use science and society. Auburn Univ Cent For Sustain, Atlanta, pp 275–278

Faldi G (2011) Evaluation of the vulnerability to climate change of the coastal communities in Dar es Salaam (Tanzania) as regards salt water intrusion in the aquifer. MSc in Environmental Engineering, Sapienza University of Rome Rome

Füssel H-M, Klein R (2006) Climate change vulnerability assessments: an evolution of conceptual thinking. Clim Change 75(3):301–329

Gallopin G (2006) Linkages between vulnerability, resilience and adaptive capacity. Glob Environ Change 16(3):293–303

IPCC (2007a) Impacts, Adaptation and Vulnerability Contribution of Working Group II to the Fourth Assessment Report of the Intergovernmental Panel on Climate Change. In: Parry ML, Canziani OF, Palutikof JP, van der Linden PJ, Hanson CE (eds) Cambridge University Press, Cambridge

IPCC (2007b) Fourth Assessment Report: Climate Change 2007 Contribution of Working Group I to the Fourth Assessment Report of the Intergovernmental Panel of Climate Change. In: Qin SD, Manning M, Chen Z, Marquis M, Averyt KB, Tignor M, Miller HL (eds) Cambridge University Press, Cambridge

Kironde JM (2001) Peri-urban land tenure, planning and regularisation: case study of Dar es Salaam. Tanzania Harare, Zimbabwe: Study carried out for the Municipal Development Programme

Kironde JM (2003a) The new land act and its possible impacts on urban land markets in Tanzania

Kironde JM (2003b) Current changes in customary/traditional land delivery systems in sub-saharan African Cities: the case of Dar Es Salaam City. University College of Lands and Architectural Studies, Dar es Salaam

Kironde J (2004) Current changes in customary/traditional land delivery systems in sub Saharan African cities: the case of Dar es Salaam city. Appendix to The Research Report R8252 Housing The Poor Through African Neo-Customary Land Delivery Systems, http://r4d.dfid. gov.uk/PDF/Outputs/Mis_SPC/R8252appendix.pdf

Kironde JM (2006) The regulatory framework, unplanned development and urban poverty: findings from Dar es Salaam, Tanzania. Land Use Policy 23(4):460–472

Kombe WJ (2005) Land use dynamics in peri-urban areas and their implications on the urban growth and form: the case of Dar es Salaam, Tanzania. Habitat Int 29(1):113–135

Lupala A (2002) Peri-urban land management in rapidly growing cities, the case of Dar es Salaam. PhD Dissertation, University of Dortmund

Marshall N, Marshall P, Tamelander JO-K, Cinner J (2010) A framework for social adaptation to climate change; sustaining tropical coastal communities and industries IUCN, Gland, Switzerland

McSweeney C, New ML (2008) UNDP climate change country profiles: Tanzania. Available at http://www.geog.ox.ac.uk/research/climate/projects/undp-cp/UNDP_reports/Tanzania/ Tanzania.lowres.report.pdf. Accessed 24 June 2013

Moser C, Satterthwaite D (2008) Towards pro-poor adaptation to climate change in the urban centers of low-and middle-income countries climate change and cities. Discussion Paper 3 IIED

Quan J, Dyer N (2008) Climate change and land tenure: the implications of climate change for land tenure and land policy FAO, Land tenure working paper 2 IIED/University of Greenwich/FAO

Ricci L (2011) Reinterpreting Sub-Saharan cities through the concept of *adaptive capacity* An analysis of *autonomous* adaptation practices to environmental changes in peri-urban areas. PhD Thesis, Sapienza University of Rome, Rome

Satterthwaite DH, Pelling M, Reid A, Romero-Lankao P (2007) Building climate change resilience in urban areas and among urban populations in low- and middle-income nations. IIED Research report commissioned by the Rockfeller Foundation

Simon D (2008) Urban environments: issues on the peri-urban fringe. Ann Rev Environ Resour 33:167–185

Simon D (2010) The challenges of global environmental change for urban. Africa Urban Forum, pp 235–248

Simone AM (2004) People as infrastructure: intersecting fragments in Johannesburg. Public Cult 16(3):407–429

Smit B, Wandel J (2006) Adaptation, adaptive capacity and vulnerability. Glob Environ Change 16(3):282–292

Tacoli C (1998) Rural-urban interactions; a guide to the literature. Environ Urban 10(1):147–166

UN (2010) World Urbanization Prospects: The 2009 Revision CD-ROM Edition—Data in digital form (POP/DB/WUP/Rev 2009), Department of Economic and Social Affairs, Population Division, New York

URT (2011) The Dar es Salaam city environment outlook, Draft Dar es Salaam: Vice President's Office, Division of Environment

Chapter 7
Mainstreaming Adaptation into Urban Development and Environmental Management Planning: A Literature Review and Lessons from Tanzania

Silvia Macchi and Liana Ricci

Abstract Mainstreaming of adaptation to climate change is recommended by many international agencies and authors of climate change literature in order to guarantee more efficient use of financial and human resources than occurs when adaptation is designed, implemented, and managed as a series of stand-alone activities. Nevertheless, there is ongoing debate over how to proceed in order to achieve effective mainstreaming, at what level to act, on what topics to concentrate, and what type of initiative should be prioritized. The article offers information and arguments that may help the administrations of Sub-Saharan cities implement the options that are most appropriate for their specific conditions. The first section situates the mainstreaming of adaptation to climate change within the more general and consolidated strategy of the Environment Integration Policy, and outlines the current process of conceptualizing the question. The mainstreaming approach is compared to the action-specific approach in order to better highlight its strengths and weaknesses, while the risks of ineffective mainstreaming are explored with particular reference to the case of Tanzania. A more detailed examination follows of possible topics and the approaches used for mainstreaming in sectoral policies related to urban development and environmental management. Lastly, the specificity of Sub-Saharan cities is addressed, which raises both concerns and hopes for the current advantages of pursuing adaptation through mainstreaming at the local level.

S. Macchi (✉) · L. Ricci
Department of Civil, Building and Environmental Engineering,
Sapienza University of Rome, Via Eudossiana 18, 00184 Rome, Italy
e-mail: silvia.macchi@uniroma1.it

L. Ricci
e-mail: liana.ricci@uniroma1.it

S. Macchi and M. Tiepolo (eds.), *Climate Change Vulnerability*
in Southern African Cities, Springer Climate, DOI: 10.1007/978-3-319-00672-7_7,
© Springer International Publishing Switzerland 2014

Keywords Adaptation mainstreaming · Literature review · Environment integration policy · Urban development planning · Tanzania

7.1 Towards Mainstreaming Adaptation in Sub-Saharan Cities

The objective of this study is to identify the basic notions and principles that could serve as references when developing a methodology for integrating climate change concerns into the urban development and environmental management plans already in place in Sub-Saharan cities.

Given the mass of conceptual frameworks and operational guidelines already available on mainstreaming adaptation to climate change and related topics, a literature review was conducted in order to explore three main questions: (1) How to define mainstreaming; (2) How to achieve it; and (3) What issues to mainstream into urban development and environmental management planning.

The debate around those three questions appears to be deeply grounded in the history of Environmental Policy Integration (EPI) and fuelled by evidence from environmental mainstreaming. Indeed, the concept of adaptation mainstreaming derives from the climate policy integration principle introduced in 2009 as an extension of the EPI principle, which traces back to the 1987 Brundtland report and related 1992 Rio Earth Summit on sustainable development. The issue of integrating environment and development in decision-making is specifically addressed by Chap. 8 of Agenda 21, adopted at the Rio Earth Summit. Governments are requested to review and improve the decision-making processes in order to ensure the integration of economic, social, and environmental considerations at all levels and in all policy sectors (UNCED 1992).

During the 2000s, environmental mainstreaming acquired an increasingly relevant role in the global strategy to fight poverty, as sound environmental management came to be acknowledged as crucial for achieving the MDGs (De Coninck 2009). In this context, risks associated with climate change and variability are viewed as a threat to the development strategies and measures in place since they "change the contexts in which development occurs", while environmental mainstreaming is the tool to ensure that development decision-making incorporates those risks, thus becoming more sustainable and successful (EC 2011).

It is therefore not surprising that, in Sub-Saharan Africa, the mainstreaming of climate change adaptation into national and sub-national development planning is usually considered part of the broader poverty-environment mainstreaming strategy, with the majority of reports and guidelines drawing from this policy background. This is particularly true in the case of Tanzania, one of the Sub-Saharan countries that made the greatest effort to apply the environmental mainstreaming approach as a means to achieving sustainable development (URT PMO-RALG 2007; URT 2010).

Lessons learned from the Tanzanian case helped identify two controversial topics which are at the front line of the debate about climate change adaptation in Sub-Saharan African cities: *mainstreaming* versus *action-specific approaches*, and mainstreaming *at the national* versus *the local level*. Accordingly, special attention is paid throughout the text to highlighting the advantages and disadvantages of the mainstreaming approach as compared to the action-specific approach. Arguments for and against at the local level of the mainstreaming approach for adaptation to climate change and variability are subsequently examined in the conclusions.

7.2 Conceptualizing Adaptation Mainstreaming

7.2.1 Defining Adaptation Mainstreaming

In development policy literature, the term *mainstreaming* is often used interchangeably with *integration* and *incorporation* to designate a strategy adopted to deal with a variety of issues, including environment and gender, which are acknowledged as crosscutting concerns in all policy fields and levels of decision making.[1]

With respect to adaptation mainstreaming specifically, although there is no consensus on a common definition, a review of research papers and policy documents draws attention to the following features:

- the definition of adaptation, which ranges from mere reduction of potential climate-related development risks to taking advantage of opportunities (OECD 2009);
- the overall goal, which ranges from ensuring the long-term sustainability of investments and reducing the sensitivity of development activities to present and future climate (Klein 2002; Huq et al. 2003; Agrawala 2005), to the broader aim of contributing to pro-poor economic growth, human well-being, and achievement of the MDGs (De Coninck 2011);
- the nature of the topic to be mainstreamed, which ranges from climate change impacts only, to societal vulnerabilities and climate change responses (Agrawala and van Aalst 2005);
- the need to simultaneously address both the mitigation of climate change causes and the adaptation of human systems to better cope with its effects (Lim and Spanger-Siegfried 2004) rather than working on adaptation in isolation;

---

[1] Consider as an example the definition of mainstreaming used in EuropeAid documents: "the process of systematically integrating a selected value/idea/theme into all domains of the EC development co-operation to promote specific (transposing ideas, influencing policies) as well as general development outcomes" (iQSG 2004 quoted in EC 2011: 16).

- the possible interchangeability of the mainstreaming adaptation approach and the climate risk management approach, although the two are slightly different, the former incorporating considerations of the long-term effects of climate change, while the latter focuses on current climate variability and no-regret measures (Lim and Spanger-Siegfried 2004);
- the field of applicability, from strictly climate change related policy sectors such as water management, disaster preparedness or land-use planning (Agrawala and van Aalst 2005), to broader areas such as "policy-making, budgeting, implementation and monitoring processes at national, sector and subnational levels" (De Coninck 2011: 3);
- the type of actors to be involved and their role within the process, based on the assumption that adaptation mainstreaming "entails working with a range of governmental and non-governmental actors, and other actors in the development field" (De Coninck 2011: 3);
- the nature of the mainstreaming process, which for the most part should be iterative rather than linear (De Coninck 2011); and
- the phasing of the process, which ranges from ex-ante determination of the objectives and measures to be integrated, to identification of adaptation objectives and measures as part of the mainstreaming process (learning process), which can lead to the reframing of planning priorities and strategies (Uittenbroek et al. 2012).

However, the best way to delineate the meaning of adaptation mainstreaming, to understand what this term includes and what it does not, is to compare it to its most common rival approach: adaptation as a dedicated domain (or the *action-specific* approach). It must be noted that such rivalry is more specious than real, as the two approaches are often considered complementary and used in combination,[2] where either special actions are viewed as generated within the mainstreaming process, or mainstreaming is viewed as a means for creating more favorable conditions for the implementation of a previously determined special action.

7.2.2 Mainstreaming Versus Action-Specific Approaches

In climate change literature, adaptation, unlike mitigation, is considered better managed through use of mainstreaming than through self-standing measures.

---

[2] An example of this is the UNDP-UNEP framework for mainstreaming climate change adaptation, which defines three levels of intervention: (i) strengthening the base for adaptation, by consciously aiming development efforts at reducing vulnerability; (ii) promoting mainstreaming adaptation measures, thus ensuring that climate change is considered in the decision-making of relevant government agencies; and (iii) promoting specific adaptation measures that target issues the first two levels have not yet tackled (De Coninck 2011). Though the mainstreaming approach is given higher priority, this framework does not exclude the possibility of recourse to special actions when satisfying results are not achieved through mainstreaming intervention modalities.

Table 7.1 Comparing action-specific and mainstreaming approaches

Action-specific approach	Mainstreaming approach
One self-standing measure	Various integrated measures
One policy sector concerned	Multi-sectoral
Few actors involved	Many actors involved
Specific knowledge and competences required	Diverse knowledge and competences required
Linear decisional process	Iterative decisional process
Replicable in many places	Context specific
Achievable in short term	Achievable in medium-long term
Standardized design and implementation	Experimental design and implementation
Clearly ex-ante defined inputs and outputs	Likely variations of inputs and outputs
Plan-driven, conformative	Target-driven, performative

Indeed, as stated by Persson and Klein (2008: 4), "adaptation is the result of a very diverse set of actions that are in turn stimulated by policy influences originating from many different sectors". The principal reason for this affirmation derives from the intrinsic complexity of the relationships between environmental changes and human systems, and the consequently high level of uncertainty in forecasting (Adger and Vincent 2005; also see Chap. 1). Moreover, the most likely futures may exacerbate current inequalities (O'Brien et al. 2012), which entails the need to consider power relations across policy sectors and levels and to involve all social actors in the development of comprehensive, multidimensional normative scenarios so that decision-making is informed by a shared vision of the future to be created.

Studies of other crosscutting development issues indicate that the mainstreaming approach, unlike the action-specific approach, has the potential for multisectoral action, and can facilitate the involvement of all levels of governance as well as a broader range of stakeholders into decision-making. Table 7.1 outlines the comparison between the two approaches.

The benefits of integrating adaptation policies and measures into development and sectoral decision-making have been identified as follows:

- To ensure consistency and avoid conflicts with other policy domains;
- To find synergies with other well-established programs, particularly sustainable development planning (Adger et al. 2007);
- To prevent *maladaptation*, or "adaptation that does not succeed in reducing vulnerability but increases it instead" (IPCC 2001: 990);
- To ensure long-term sustainability (Persson and Klein 2008);
- To ensure that future projects and strategies will reduce vulnerability by including priorities that are critical to successful adaptation (Lasco et al. 2009);
- To reduce the sensitivity of development outcomes to current and future climate change and variability (Huq et al. 2003; Agrawala 2005; Klein et al. 2005, 2007);
- To make more sustainable, efficient, and effective use of financial and human resources than is achieved when adaptation is designed, implemented, and managed separately from development (Persson and Klein 2008);

- To leverage the much larger financial flows in sectors affected by climate risks than the amounts available for financing adaptation separately (Agrawala 2005); and
- To enhance the performance and development contribution of each sector and each government body at all levels.

Nevertheless, there are many difficulties in applying the mainstreaming approach, for which its actual effectiveness is often questioned. Some of those difficulties do not arise in the case of the action-specific approach, which leads governments to prefer the latter over mainstreaming. In fact, the action-specific approach seems to be simpler in terms of planning, acquisition of financing and other necessary means, decision-making, implementation, and evaluation of results. In addition, it appears that the transformative potential of the main-streaming approach can be seriously compromised by a lack of shared, short-term goals, which can reduce the motivation of and push from public administrations as regards changing norms, procedures, and organizational models. Therefore, the overall hope of success for a special action is certainly higher than that of a mainstreaming initiative but, given its characteristics, the efficacy of a special action as regards reducing vulnerability in the medium-long term with respect to risks that at present are not entirely knowable is not at all guaranteed. The problem then arises of how to avoid the risk of ineffective mainstreaming. The extent to which the case of Tanzania can be enlightening in this respect is discussed below, following a description of the methods and tools for achieving mainstreaming.

7.3 Operationalizing Mainstreaming

7.3.1 How to Achieve Mainstreaming

In the EPI literature, methods and tools for integration are usually grouped into four operational streams, which are neither mutually exclusive nor exhaustive: *procedural, organizational, normative*, and *reframing* approaches (Persson 2007; Persson and Klein 2008).

The *procedural approach* consists of introducing new or modifying existing decision-making procedures while feeding information related to the issue to be mainstreamed into decisional processes. Procedural tools commonly used in environmental mainstreaming include ex ante environmental assessments of pro-grams and projects, green budgeting and checklists, sectoral environmental reporting and audits, consultation with environmental experts, and participation of stakeholders.

It should be noted that the purpose of these changes is not to target specific decisions, but to contribute to changing the overall context in which decision-making occurs, which in turn will reorient all subsequent decision-making pro-cesses (Lenschow 2002 cited in Persson and Klein 2008).

In order to be effective, procedural changes must be defined and applied simultaneously at the various levels of government involved directly or indirectly in the decision-making process. Otherwise, their efficacy runs the risk of being compromised by the lack of knowledge, awareness, and resources for implementation at any of the levels involved.

The *organizational approach* involves amendments to formal responsibilities and mandates, creating new or merging existing institutions, networking among diverse departments, and structural changes of budgets. It intervenes at every level of organization, from the individual to the general one, to induce ownership, appropriation, understanding, and enhanced capabilities on the issue to be mainstreamed within the relevant sectors as well as to introduce new responsibilities and various accountability mechanisms (Peters 1998 cited in Persson and Klein 2008).

Institutional inertia, sectoral compartmentalization, self-interest, and related budget competition might jeopardize effective organizational changes. On the other hand, this type of change plays a fundamental role in increasing the effectiveness and efficiency of the policy-making processes, and as such should be considered "a general principle for good decision making" (Nilsson and Persson 2003: 335) rather than a specific response to the requirements of mainstreaming.

The organizational approach is suitable for use at all government levels and, among the four approaches considered here, is likely the most applicable at the local level as it does not necessarily involve similar changes at upper levels.

The *normative approach* implies a change in policy-making culture. As such, it requires high-level commitments to the issue to be mainstreamed and entails the formalization of that issue in existing strategies and policy frameworks as well as the allocation of additional targeted resources. This approach usually leads to "commitments to particular goals in treaties and directives, requirements for sectoral strategies, obligations to report performance, and external and independent reviews" (Persson and Klein 2008: 10).

The *reframing approach* aims at reshaping the policy frame of traditional sectors in a mid- and long-term perspective. It helps in integrating the issue to be mainstreamed into the perception of the function and objectives of a sector among relevant stakeholders (e.g. by relabeling policy fields as *adaptive* health policy or *adaptive* land-use policy). While this is a lengthier process than an operational approach, it can be stimulated through research, training, and diversification of staff in terms of backgrounds and cultures (ibid.).

Both the normative and the reframing approaches are more appropriately applied at the government levels, usually national and international, where laws are elaborated and policies formulated. Nevertheless, these approaches also have important implications at the levels of government where new norms are applied and new policies are implemented, as they require procedural and organizational changes.

7.3.2 Risks of Ineffective Mainstreaming: The Case of Tanzania

The environmental mainstreaming initiated in Tanzania over the last two decades has stimulated the decentralization of decision-making and administration, which were extremely slow despite a number of reforms deliberated by the parliament (Lerise 2000). At the same time, the central government has viewed environmental mainstreaming as a useful tool for maintaining control of local governments (Death 2013).

On the other hand, evaluations of environmental mainstreaming are less positive with respect to the reduction of poverty, food insecurity, and ecological degradation. Results do not appear to be on par with the level of financial resources (contributed predominantly by donors) that the Tanzanian government has invested in the reform of environmental governance since the 1990s'.

> Allegations of chronic mismanagement, rent-seeking and pervasive corruption in the forestry, fisheries and wildlife conservation sectors further contribute to substantial doubts over the efficacy of the much-lauded environmental mainstreaming (Death 2013: 3).

Death (ibid.) recognizes that post-sovereign environmental governance in Tanzania is non-exclusive, non-hierarchical, and non-territorial. Nevertheless, such characteristics do not guarantee positive outcomes. Firstly, the involvement and inclusion of a multiplicity of actors (one example is the mainstreaming of environmental concerns in the MKUKUTA—URT 2005 and 2010) may not always result in improved legitimacy, accountability, and transparency, due to power imbalances and competition over control of resources. Secondly, the old hierarchies have not been replaced by a coherent framework of local and national scale competencies and levels of authority. The resulting ambiguity of decision-making and administrative processes, together with the lack of transparency in business relationships and practices, guarantees that the system only works for those who have power and a capacity to negotiate that shields them from the pressures exerted by other actors. Thirdly, although the Tanzanian government has adopted a problem driven and transboundary (or post-territorial) logic in structuring environmental governance (e.g. protection initiatives for 'global' biodiversity through transboundary conservation areas), there is no shortage of territorially bounded initiatives informed by a nationalist discourse. Death (ibid: 19–22) maintains that "environmental planning and management in Tanzania can be seen in terms of a longer dynamic set of power relations between the central state bureaucracy and the rural population". From this perspective, several instruments typical of environmental governance—including monitoring, evaluation, auditing, and surveillance over the natural environment—may mean, in practice, greater state control and penetration over rural populations.[3]

---

[3] Death (2013) cites two cases in support of his thesis: the Kilimo Kwanza agricultural development policy and the National Strategy for Growth and Reduction of Poverty MKUKUTA

Under such circumstances, the mainstreaming process becomes inevitably ineffective if not counterproductive. The opacity and/or dynamicity of roles and relations make it difficult to identify entry points for initiating processes of change or the actors responsible for monitoring those changes. It is also difficult to plan and monitor the distribution of resources and relationships between the investment of resources and results achieved. Moreover, environmental analysis and planning, like certain normative or procedural changes (e.g. the mandatory EIA law) can become instruments of spatial control exercised by the state. Lastly, mainstreaming can become a method for de-territorializing the distribution of funds, with the risk that resources will become volatilized and distributed in an unbalanced manner among various areas and actors.

The case of Tanzania is emblematic as regards two questions that are thoroughly explored in the literature. From an institutional perspective (Brouwer et al. 2013), the efficacy of mainstreaming seems to require the presence of two specific conditions: (1) regulatory capacity of public authorities, and (2) balance of power and resources between environmental and sector stakeholders or authorities. Capacity is directly linked to resources (both financial and competence-related) legitimization, and information. If resources and capacity are lacking, mainstreaming is compromised.

A notable critique of mainstreaming as a strategy that draws on the potential for change within bureaucratic organizations has been offered by Longwe in her writing on gender mainstreaming (1997). That author maintains that, alongside the overt bureaucracy within an organization, there exists another covert bureaucracy that is capable of subverting all policies and directives that threaten covert patriarchal interests.

> The overt organization is a conventional bureaucracy, which is obliged to implement policies handed down by government. The covert organization is what we have here called the 'patriarchal pot', which ensures that patriarchal interests are preserved. When presented with feminist policies, the overt and the covert organizations have opposing interests, values, rules and objectives: bureaucratic principles demand implementation, while patriarchal principles demand evaporation (ibid.).

Similarly, mainstreaming of environmental or adaptation concerns is destined to fail when implementation involves changes that threaten value systems and interests that the bureaucracy tends to perpetuate.

(Footnote 3 continued)
I and II (URT 2005 and 2010). Regarding the latter in particular, he states that "the repeated stress on the very existence of quantitative and measurable targets and indicators—rather than success or failure in meeting them – is evidence of this focus [on statist control over land and territory]" (ibid.: 22).

7.4 Mainstreaming What into What

7.4.1 Climate Proofing Versus Adaptive Capacity Improvement

The literature on adaptation mainstreaming can be divided into two different streams depending on how vulnerability is interpreted: in one case, vulnerability is viewed as a linear result of climate change impacts on an exposure unit (*outcome vulnerability*), while in the other, climate change is considered to interact dynamically with contextual conditions associated with an exposure unit (*contextual vulnerability*) (O'Brien et al. 2007).

Approaches from an outcome vulnerability perspective seek to limit negative outcomes of climate change by securing the physical environment, especially the city, through improved infrastructure and measures for impact mitigation. The ultimate goal is the climate proofing of policies and plans, or of development in general. This is often described as mainstreaming disaster-risk reduction in development planning (Khailania and Pererai 2013) or mainstreaming climate change adaptation into comprehensive disaster management. Within this stream, screening is undertaken to establish relevance to climate change effects and to justify further examination of climate risks, and it is complemented by a risk assessment consisting of a detailed examination of the nature of climate risk and of possible risk management strategies.

Approaches from a contextual vulnerability perspective address the issue of human security in a more multidimensional manner, rather than considering it as merely achievable through a secured physical environment. They focus on the improvement of adaptive capacity (see Chaps. 1 and 6), drawing upon the sustainability livelihood framework and Sen's capability approach to development. At the center of this reasoning is the idea that vulnerability is the result of a process in which the system of social interactions and power relations influences people's access to resources, and therefore contributes in a determinative manner to defining the kind of vulnerability of a given social group in a given time and place. As a result, the adaptation measures defined through such an approach also address the structural inequalities of the context in order to change vulnerability circumstances (O'Brien et al. 2007; Simon 2010).

Important indications for reciprocal learning and general improvement of the efficacy of both streams are provided by the IPCC Special Report (2012).[4] In particular, the report recognizes that assessment must also consider a series of non-climate factors that modify contextual conditions with notable consequences in

---

[4] It should be noticed that, to date, major development agencies have not make a clear choice between the two approaches. The already mentioned UNDP-UNEP framework (De Coninck 2011), for instance, while paying particular attention to addressing the adaptation deficit and increasing the overall resilience of the country and population, does not exclude the need for climate-proofing policies.

terms of both determining climate change impacts and shaping people's adaptive capacity.

7.4.2 Mainstreaming Adaptation into Urban Development and Environmental Management Sectors

Despite the fact that there is widespread consensus on the necessity of mainstreaming adaptation into sectoral plans and programs at the local level, most adaptation mainstreaming research and practices have focused on development policy at the national level (Klein 2002; Huq et al. 2003; Agrawala 2005; Persson and Klein 2008). This provides a valuable theoretical and practical basis for advancing at the sectoral level, though each individual sector requires a specific effort according to its own cultural tradition and the network of interests that it mobilizes.

Urban development planning and the management of environmental problems in cities stand out as two sectors that tend to evolve from the local level rather than the national one, due to their territorial specificity. At present, attention seems to be focused on the procedural and organizational levels, for example through the preparation of adaptation action plans and the creation of dedicated task forces, in a manner that is not very different from the implementation of the Rio Agenda 21. Meanwhile, the process of change for normative aspects still seems to be at the embryonic stage, and policy reframing appears even less advanced.

There are several reasons why these sectors should be considered the front line in the adaptation mainstreaming process. Some of those reasons derive from the very nature of the city, others from the potential for adaptation intervention and the experience accumulated regarding environmental mainstreaming in these sectors. Moreover, as concerns Sub-Saharan cities, the coupling of climate change and rapid urbanization places increased strain on the urban planning and management capacity of local authorities (O'Brien et al. 2012).

Firstly, a large number of potential climate change victims is concentrated in cities, as are fundamental assets and activities for the production of a large part of national wealth. At the same time, urban development is both one of the main causes of climate change and one of the non-climate factors destined to exacerbate the effects of climate change. In addition, cities hold a concentration of capacities and resources for innovation, and are therefore privileged places for the elaboration of possible solutions to the emerging challenges, i.e. adaptation.

As regards the potential of intervention, it should be noted that urbanization itself is not always a driver of increased vulnerability. Instead, the type of urbanization and the context in which urbanization is embedded defines whether these processes contribute to an increase or decrease in people's vulnerability. The ability to carry out urban planning in an effective way is part of local capacity for adaptation, but it needs time to produce significant effects (Cardona et al. 2012). There is widespread consensus that land-use planning and ecosystems management have beneficial effects in terms of providing environmental services that are crucial to supporting people's livelihoods as well as disaster risk protection services for infrastructure, water resource management, and food security (Lal et al. 2012). Furthermore, urban planning has the potential to create synergies between climate change adaptation and mitigation measures, while the importance that the choice of one urban form over another can have in terms of improving adaptive capacity and the reduction of GHG emissions is more controversial (O'Brien et al. 2012).

The most common approaches for mainstreaming adaptation in urban development and environmental management draw upon the well-established field of sustainable urban development (Cohen et al. 1998) and the more recently developed field of urban resilience (Pelling 2003; Davoudi and Porter 2012). Work streams can be distinguished that concentrate on various aspects of the urban reality and require the contributions of three different disciplinary groups: applied technological and infrastructure-based approaches; human development and vulnerability reduction; and investing in natural capital and ecosystem-based adaptation. The IPCC (Lal et al. 2012) suggests combining the contributions of these streams, since all three address complementary and useful aspects for effective adaptation.

In fact, what is most important is reinforcing the capacity of administrations to manage the uncertainty surrounding future changes and to uncover issues of justice and fairness embedded in the procedures for decision-making and the distribution of burdens and benefits.

As concerns the first point, the literature on adaptation (ibid.) converges on the following points: (i) investment in improved knowledge of local climate change effects; (ii) integration of available information into decisions; (iii) in the absence of robust information, consideration of *no or low regrets* strategies; and (iv) preference for reversible interventions and flexible decision-making processes in order to allow for ongoing adjustment as new information becomes available.

The second point recalls the broadly discussed question *planning for whom?* Although addressing such a question is beyond the limits of the present text, it must be noted that any effort to adapt to climate change is destined to fail if it does not take into account the fact that any policy has differential impacts across temporal and spatial scales as well as social groups. As remarked by the IPCC (Cutter et al. 2012: 320), there is an important gap in adaptation research as regards "the mechanisms or practical actions needed for advancing social and environmental justice at the local scale, independent of the larger issues of accountability and governance at all scales".

7.5 Concerns and Hopes for Adaptation Mainstreaming in Sub-Saharan African Cities

A variety of lines of reasoning converge on the importance of the local dimension in determining the efficacy of adaptation mainstreaming in urban development and environmental management policy and planning.

First, it is fairly obvious that the "culture of planning" of a particular place will heavily influence the possibility of practicing the mainstreaming approach as opposed to engaging in special actions, and will determine the content of adaptation. Friedman (2005) sought to typify urban planning in Africa on the basis of a few characteristics that the cities of the continent have in common: average urban growth of at least 5% annually; implosion of the informal economy upon which the urban poor depend for their survival; financial incapability of adequately servicing the population; and allocation of the majority of the land without regard for regulations and planning standards. It is therefore appropriate to ask whether it makes sense to proceed with the mainstreaming of adaptation within the context of urban plans that are usually in default of implementation.

Nevertheless, mainstreaming across different sectors (horizontal linkages), which is fundamental when addressing urban issues, is certainly more practical at the local level since conflicts between competing priorities are more evident and shared interests in avoiding socio-ecological crises are stronger among actors who co-habit the same place. From this perspective, it is essential that the implementation of mainstreaming includes participatory practices capable of involving the population that most depends on natural resources and is therefore most likely to suffer from the effects of climate change.

However, sectoral mainstreaming at the national level is necessary in order to create a favorable context for the sustainability and up-scaling of successful local level practices (vertical linkages). In this respect, the local dimension can also represent a formidable resource in terms of guaranteeing that mainstreaming occurs from the bottom-up. The conflict between national interests in the city as a motor of economic growth and the interests of the majority of the urban population, which sees the city as a resource for carrying out their own life plans, is inevitable and particularly relevant in Eastern Africa. However, the sharing of a common space renders negotiation between the two parties of reciprocal value and to a certain extent facilitates the identification of win–win solutions.

Lastly, there is an important obstacle to the adoption of the mainstreaming approach to adaptation to climate change and variability in least developed countries. The concern in these countries is that the choice of mainstreaming implies a reduction of the funds dedicated to adaptation (Michaelowa and Michaelowa 2007) or their absorption within development funding and the risk that they will be directed to other objectives (Yamin 2005). In addition, there is concern that donors may use adaptation mainstreaming to impose certain conditions (Gupta and van de Grijp 2010). This obstacle also arises between national and local governments, particularly under conditions like those mentioned in Sect. 7.3.2.

In conclusion, adopting the mainstreaming approach to climate change adaptation in Sub-Saharan cities is far from a simple choice. However, in our opinion it is the best option because, notwithstanding the above mentioned difficulties and risks, mainstreaming offers a unique opportunity to reframe urban policy and free it of the paradigms of modern urban planning, allowing it to pursue promising new directions in the sphere of African urban thought (Ricci 2011).

References

Adger WN, Vincent K (2005) Uncertainty in adaptive capacity. CR Geosci 337:399–410

Adger WN, Agrawala S, Mirza MMQ, Conde C, O'Brien K, Pulhin J, Pulwarty RS, Smit B, Takahashi K (2007) Assessment of adaptation practices, options, constraints, and capacity. In: IPCC, climate change 2007: impacts, adaptation and vulnerability. Contribution of working group II to the fourth assessment report of the intergovernmental panel on climate change. Cambridge University Press, Cambridge, p 717–743

Agrawala S (ed) (2005) Bridge over troubled waters: linking climate change and development. OECD Publishing, Paris, p 154

Agrawala S, van Aalst M (2005) Bridging the gap between climate change and development. In: Agrawala S (ed) Bridge over troubled waters: linking climate change and development. OECD Publishing, Paris, pp 133–146

Brouwer S, Rayner T, Huitema D (2013) Mainstreaming climate policy: the case of climate adaptation and the implementation of EU water policy. Environ Plann C 31(1):134–153

Cardona OD, van Aalst MK, Birkmann J, Fordham M, McGregor G, Perez R, Pulwarty RS, Schipper ELF and Sinh BT (2012) Determinants of risk: exposure and vulnerability. In: IPCC (2012) op. cit., p 65–108

Cohen S, Demeritt D, Robinson J, Rothman D (1998) Climate change and sustainable development: towards dialogue. Glob Environ Change 8(4):341–371

Cutter S, Osman-Elasha B, Campbell J, Cheong S-M, McCormick S, Pulwarty R, Supratid S, Ziervogel G (2012) Managing the risks from climate extremes at the local level. In: IPCC (2012) op. cit., p 291–338

Davoudi S, Porter L (2012) Interface: applying the resilience perspective to planning: critical thoughts from theory and practice. Plann Theor Pract 13(2):299–333

De Coninck S (2009) Mainstreaming poverty-environment linkages into development planning: a handbook for practitioners. UNDP-UNEP, Nairobi, p 120

De Coninck S (2011) Mainstreaming adaptation to climate change into development planning: a guide for practitioners. UNDP-UNEP, Nairobi, p 86

Death C (2013) Environmental mainstreaming and post-sovereign governance in Tanzania. J East Afr Stud 7(1):1–20

EC (2011) Guidelines on the integration of environment and climate change in development cooperation. Tools and methods series 4:18. http://capacity4dev.ec.europa.eu. Accessed 28 July 2013

Friedmann J (2005) Globalization and the emerging culture of planning. Prog Plann 64(3):183–234

Gupta J, van de Grijp N (eds) (2010) Mainstreaming climate change in development cooperation: theory, practice and implications for the European union. Cambridge University Press, Cambridge

Huq S, Rahman A, Konate M, Sokona Y, Reid H (2003) Mainstreaming adaptation to climate change in least developed countries (LDCs). IIED, London, p 40

IPCC (2001) Climate change 2001: impacts, adaptation, and vulnerability. In: McCarthy JJ, Canziani OF, Leary NA, Dokken DJ, White KS (eds) Contribution of working group II to the third assessment report of the intergovernmental panel on climate change. Cambridge University Press, Cambridge and New York

IPCC (2012) Managing the Risks of Extreme Events and Disasters to Advance Climate Change Adaptation. In: Field CB, V Barros, TF Stocker, D Qin, DJ Dokken, KL Ebi, MD Mastrandrea, KJ Mach, G-K Plattner, SK Allen, M Tignor and PM Midgley (eds) A special report of working groups I and II of the intergovernmental panel on climate change. Cambridge University Press, Cambridge and New York

Khailania DK, Pererai R (2013) Mainstreaming disaster resilience attributes in local development plans for the adaptation to climate change induced flooding: A study based on the local plan of Shah Alam city Malaysia. Land Use Policy 30(1):615–627

Klein RJT (2002) Climate change, adaptive capacity and sustainable development. Paper presented at the expert meeting on adaptation to climate change and sustainable development, OECD, Paris, 13–14 March 2002

Klein RJT, Schipper EL, Dessai S (2005) Integrating mitigation and adaptation into climate and development policy: three research questions. Environ Sci Policy 8:579–588

Klein RTJ, Eriksen SHE, Naess LO, Hammill A, Tanner TM, Robledo C, O'Brien KL (2007) Portfolio screening to support the mainstreaming of adaptation to climate change into development assistance. Clim Change 84:23–44

Lal PN, Mitchell T, Aldunce P, Auld H, Mechler R, Miyan A, Romano LE and Zakaria S (2012) National systems for managing the risks from climate extremes and disasters. In: IPCC (2012) op. cit., p 339–392

Lasco RD, Pulhin FB, Jaranilla-Sanchez PA, Delfino RJP, Gerpacio R, Garcia K (2009) Mainstreaming adaptation in developing countries: the case of the Philippines. Clim Dev 1(2):130–146

Lerise F (2000) Urban governance and urban planning in Tanzania. In: Ngware S, Kironde JML (eds) Urbanising Tanzania: issues, initiatives, and priorities. University of Dar es Salaam Press, Dar es Salaam, p 88–116

Lim B, Spanger-Siegfried E (eds) (2004) Adaptation policy frameworks for climate change: developing strategies, policies and measures. UNDP, Cambridge University Press, Cambridge

Longwe SH (1997) The evaporation of gender policies in the patriarchal cooking pot. Dev Pract 7(2):148–156

Michaelowa A, Michaelowa K (2007) Climate or development: is ODA diverted from its original purpose? Clim Change 84(1):5–21

Nilsson M, Persson A (2003) Framework for analysing environmental policy integration. J Environ Plann Policy Manage 5(4):333–359

O'Brien K, Pelling M, Patwardhan A, Hallegatte S, Maskrey A, Oki T, Oswald-Spring U, Wilbanks T, Yanda PZ (2012) Toward a sustainable and resilient future. In: IPCC (2012) op. cit., p 437–486

O'Brien K, Eriksen SHE, Schjolden A, Nygaard LP (2007) Why different interpretations of vulnerability matter in climate change discourses. Clim Policy 7:73–88

OECD (2009) Policy guidance on integrating climate change adaptation into development co-operation. OECD Publishing, Paris. doi:dx.doi.org/10.1787/9789264054950-en

Pelling M (2003) The vulnerability of cities: natural disasters and social resilience. Earthscan, London

Persson Å (2007) Different perspectives on EPI. In: Nilsson M, Eckerberg K (eds) Environmental policy integration in practice: shaping institutions for learning. Earthscan, London, pp 25–48

Persson Å, Klein RJT (2008) Mainstreaming adaptation to climate change into official development assistance: integration of long-term climate concerns and short-term development needs. In: Proceedings of the Berlin conference on the human dimensions of global environmental change, Berlin, 22–23 Feb 2008

Ricci L (2011) Reinterpreting Sub-Saharan cities through the concept of "adaptive capacity". An analysis of "autonomous" adaptation practices to environmental changes in peri-urban areas.

PhD in Urban Planning. Sapienza University, Rome. Italian version. http://padis.uniroma1.it/handle/10805/1375. Accessed 28 July 2013

Simon D (2010) The challenges of global environmental change for urban Africa. Urban Forum 21(3):235–248

Uittenbroek CJ, Janssen-Jansen LB, Runhaar HAC (2012) Mainstreaming climate adaptation into urban planning: overcoming barriers, seizing opportunities and evaluating the results in two Dutch case studies. Reg Environ Change 13(2):399–411

UNCED Secretariat (1992) Agenda 21. United Nations, Geneva

URT (2005) National strategy for growth and reduction of poverty (MKUKUTA). Prime Minister's Office, Dodoma

URT (2007) National Framework for urban development and environmental management (UDEM) in Tanzania. Volume II: UDEM final framework design report. Prime Minister's Office—Regional Administration and Local Government, Dodoma

URT (2010) National strategy for growth and reduction of poverty II (MKUKUTA II). Ministry of Finance and Economic Affairs, Dar es Salaam

Yamin F (2005) The European union and future climate policy: is main-streaming adaptation a distraction or part of the solution? Clim Policy 5(3):349–361

Chapter 8
Knowledge Sharing on Climate Change as a Resource for Adaptation Processes: The Case of Malika, Senegal

Rita Biconne

Abstract In order to face Climate Change, cities should have a complex system of adaptation capacities and abilities. Cities should involve everyone in this process, disseminate information, and spread awareness of Climate Change issues. Conventional studies and approaches are based on scientific knowledge that rarely takes into consideration the socio-cultural aspects and practices prevalent in specific contexts. The assumption on which this chapter is based, however, is that inclusion of these components is important and even necessary to starting a real process of adaptation. Therefore, we consider adaptation as a continuum in which strategy development depends on the relationships between local authorities, the private sector, researchers, and civil society. In this context, building adaptive capacity, knowledge sharing, and the delivery of adaptation actions become phases of the methodology. Malika, a peri-urban area of Dakar in Senegal, has been vulnerable to flooding since 2005, undermining dwellers' daily lives and productive activities, particularly agriculture. This chapter aims to highlight how the sharing of knowledge on Climate Change, through a participatory approach, can be a useful tool in the decision-making processes that characterize urban planning.

Keywords Adaptation planning · Knowledge sharing · Participatory approach · Peri-urban Africa · Senegal

R. Biconne (✉)
Urban and Regional Planning Department, University of Florence, Via Micheli 2, 50121 Florence, Italy
e-mail: rita.biconne@unifi.it

S. Macchi and M. Tiepolo (eds.), *Climate Change Vulnerability in Southern African Cities*, Springer Climate, DOI: 10.1007/978-3-319-00672-7_8, © Springer International Publishing Switzerland 2014

8.1 Introduction

The present research adopts the definition of adaptation as adjustments to enhance the viability of social and economic activities and to reduce their vulnerability to climate, including its current variability and extreme events as well as longer term climate change (Smit et al. 2000).

As such, adaptation is any alteration in the state of a system, in response to a stressor, under which key variables are conserved or enhanced. This systems definition of adaptation directs attention towards uncovering processes rather than accounting for specific events or resources. Considering adaptation in terms of learning allows for the identification of both material adaptation, individual and collective actions, and institutional modification as valid adaptive strategies (Pelling et al. 2008). In this framework, knowledge and learning as forms of adaptive behavior raise questions about the process through which actors can share and can learn to be adaptive.

Although the importance of the vertical interaction between local, regional, and national actors is well established, focus on the local level, where adaptive behavior is most prominent, is often inadequate. This chapter argues that the relational attributes of local actors, their degree of knowledge, and social implications are central to adaptive capacity, which enables responses not only to the unforeseen shocks related to climate change, but also to the uncertainty of economic, social, and political change.

The following sections aim to highlight the relevance of knowledge sharing through the use of a participatory approach as a tool in the decision-making processes of urban planning in Malika, Senegal. Therefore, the main objectives of this contribution are:

1. to assess the degree of awareness and perception of local urban issues, in terms of both generalized and more specifically regional impacts of climate change;
2. to demonstrate the application of a participatory approach to sharing knowledge as a potential tool in the decision-making processes of urban planning, based on a case study of Malika; and
3. to make recommendations regarding the importance of collaborations among various urban actors when establishing local strategies and measures for adaptation to Climate Change (CC).

8.2 From the Approaches of Classical Science to Knowledge Sharing

Natural disasters and increasing environmental stresses are not only capturing the attention of direct victims of specific natural disasters and niche specialists. The challenge of climate change is more and more intensely affecting the international scientific debate, political and institutional channels, and public opinion.

In the last decade, the production of information on impacts, vulnerability, and adaptation to climate change, has adopted the classic approach based on scientific recommendations. The tendency of politicians to turn to experts, often abroad, to define a strategic framework has certainly created knowledge of considerable value, but it has been less effective for local planning.

It is increasingly recognized that adaptation approaches, rooted in part in indigenous knowledge and its ability to survive, are likely to be less efficacious when they are imposed through top-down initiatives. At the same time, it is accepted that, given the pace of transformation and the increasing severity of CC impacts, indigenous practices are not sufficiently autonomous in the adaptation process.

In this perspective, a close collaboration between local and expert knowledge, including innovations in science and indigenous practices, is a fundamental interaction.

Historically, the creation and dissemination of knowledge was the monopoly of certain individuals or institutions. This resulted in the marginalization of segments of society based on gender, race, language, and other discriminating factors. However, the emergence of participatory tools has led many to argue that a new *architecture of participation* is emerging that will democratize access to and production of knowledge (Thompson 2008).

A recent report indicates the tensions and mistrusts in the relations among African scientists, journalists, and politicians as the main obstacle to open exchanges (Ochieng 2009), and emphasizes the need for an improved level of mediation among these actors and reinforced cooperation aimed at an additional increase of knowledge.

The urgent need to address the challenge of adaptation is closely linked to a growing recognition that the success of adaptation practices has to take account of local components (Dinar et al. 2008, Maddison 2006) and the collaboration with local institutions and local social actors (Agrawal and Perrin 2008).

Recognizing, also, the potential of indigenous knowledge (which is often transmitted orally at the local scale and not formally documented) in the adaptation process is important when determining appropriate modes of interaction among the holders of this immaterial common good.

Even international bodies are moving in this direction. According to the United Nations Framework Convention on Climate Change (UNFCCC 2007), every national government should prepare for and facilitate adequate adaptation to climate change, including the adjustments necessary at the community, national, and international levels. It has been stressed that adaptation strategies must include an appropriate mutual relationship between the use of suitable technologies and information on traditional coping practices, and they should also include a diversification of current livelihoods and local interventions.

A detailed analysis by the International Research Institute for Climate and Society, based on a *gap analysis* of the level of information on CC, indicates that there is a paucity of scientific evidence of the value of integrating such information in decision-making. That research also emphasizes the need to engage all stakeholders, to facilitate awareness and education, and to support dialogues so that users can help to shape the services they receive (Hellmuth et al. 2007: 10).

These are the assumptions of the analytical and methodological research approach illustrated below.

8.3 Methodology

The research began in 2008 as the Masters thesis of the present author and another Italian architect, Loredana Lucentini. It is partially integrated into the doctoral research of the author on the value of water in regional land-use and local transformation. That study highlights the settlement dynamics in Dakar's metropolitan area, and specifically examined the effects of flooding in the urban context of Malika. The overall aim was to provide municipalities, local associations, and traditional authorities with updated tools and documentation, which were shared or created in collaboration with inhabitants as a support for decision-making with respect to territorial interventions.

The research was divided into two phases. The first phase involved a literature review and cartographic research, data collection, and historical settlement analysis; the second phase included a four-month mission organized with the support of the Italian NGO Fratelli dell'Uomo and conducted in collaboration with the Senegalese associations *Intermondes* and CREM (*Centre de Ressources Educationnelles Malika*).

8.3.1 Introduction of Case Study

Senegal has been characterized by unfavorable socio-economic conditions for decades. The country is ranked 156th on the United Nations Human Development Index, and more than 56 % of the population lives below the poverty line.

In Senegal, as in most African countries, urbanization and settlement processes have mainly been characterized by top-down interventions, the result of colonial rule and inappropriate codes of modern land management. For example, the Dakar *Plan d'Urbanisme*, in use since 1967—the year immediately following independence—demonstrates the inadequacy of planning tools when faced with the rapidity of urban transformations and pressing environmental emergencies.

Since the '1980s, increasing portions of the Senegalese territory have been subject to the worsening of environmental and climatic conditions: the northern part in particular has suffered severe drought, almost the entire coastline is vulnerable to rising sea levels, and various urban areas in the region of Dakar have been subject to devastating floods.

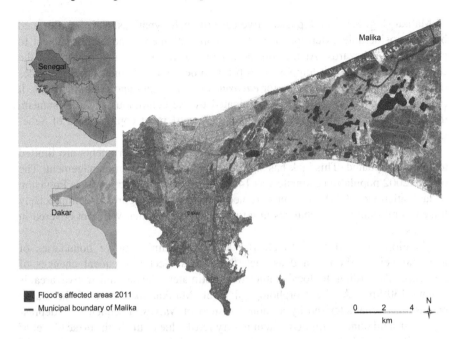

Fig. 8.1 Location of Malika in the Dakar region

Vulnerability to flooding is often related to a lack of urban services and infrastructure, such as drainage systems, and the quality of settlements.

This is worth considering more generally in that so much of the human cost of extreme weather events in urban centres in low- and middle-income nations comes not from the *hazard* or the *disaster-event* but from the inadequacies in provisions to protect urban populations (or particular sections of the population) from it (Satterthwaite et al. 2007: 9).

In this framework, ecologically fragile areas are mainly suburbs and zones occupied by informal settlements, where people can still use and occupy land for subsistence (ActionAid International 2006). These inhabitants, often economically and socially excluded, are frequently the part of the population that is most vulnerable to the impacts of climate changes.

Some people migrate to urban centers in search of better employment and better economic opportunities—known as pull factors—while others migrate to escape the negative effects of climate change, such as drought, floods, hunger, social inequality, and cultural and spatial policies—known as push factors.

Environmental stress is a known contributing factor to rural–urban migration and urbanization processes in Africa (Hope 2009).

Urban migrants are often subject to insufficient and unequal access to housing (Yuen and Kumssa 2010; Simone 2004; Diop 2004), which is mainly of a precarious and informal nature and located in confined and not built areas, where they are more exposed to environmental hazards.

Malika (Fig. 8.1) is a representative case of such dynamics, as it is influenced by the considerable extension of metropolitan settlements and highly subject to environmental hazards. At the threshold between the Dakar metropolis and the rural world, Malika is a temporary stop for some, though it more often becomes the permanent area of residence for *environmental refugees* and urban migrants. It represents a spatial reality that is permeated by the continuous social dynamism through which migrants manage to change the settlement configuration of Senegalese cities (Piermay and Sarr 2007).

The demographic and spatial data available to local governments are limited and poorly updated. This lack has significant impacts on urban management. The official 2002 population estimate was 14,167 inhabitants (Direction de la Prévision et la Statistique 2004); one must remember that these figures are imprecisely linked to migrations and changes in informal settlements that were not covered in the census.

Therefore, it is difficult to clearly define the administrative boundaries of marginal neighborhoods, and as such they are subject to frequent changes of influence. According to local authorities' estimates, the administrative area is around 130 km^2. At the morphological level, Malika, like most of the Dakar peninsula, is characterized by a natural system of *Niayes*, a sequence of depressions and sand dunes. This ecosystem is very fertile due to the high surface level of the aquifers, which favors agricultural production, but at the same time it is one of the areas that is most fragile and vulnerable to floods.

As regards its historical and religious foundation, Malika is characterized by the coexistence of traditionally legitimized figures and more recently established legal representatives of the municipality.[1] Changes in the distribution of powers and territorial responsibilities have radically modified existing local structures and required considerable efforts of adjustment and cooperation between the two local entities. Nevertheless, local government, invested with new and relevant assignments, has encountered many obstacles to improving land management strategies in a synoptic frame of spatial and environmental components.

8.3.2 Knowledge Cannot be a Luxury Anymore. A Participatory Process of Sharing

Behaviors at the individual and social level are linked to a variety of factors that affect decisions regarding processes of adaptation to climate change. Actions and practices are largely shaped by the cultural and social values of a given context.

---

[1] The municipality of Malika was established following the implementation of the National Law in 1996 with the transfer of competences from the State to local authorities in the main categories of decentralization, including land use planning, urban design and housing.

Fig. 8.2 Participatory approach to knowledge sharing processes

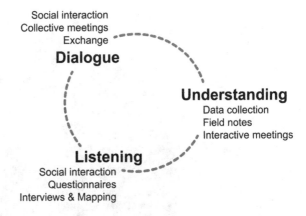

Local practices include socially constructed beliefs, preferences, and perceptions that can undergo significant changes over time.

According to the studies conducted by Cambridge University on the psychology of risk, effective appreciation of social and individual actions, and consequent limits, must be understood in relation to the characteristics of a specific context; vice versa, how those characteristics affect perceptions and related behaviors must also be understood (Adger et al. 2009).

As mentioned above, the present research identifies the various levels of social perception and awareness of climate issues as a central point (Fig. 8.2). The main stages of research on social perception were based on three forms of interaction with and between local actors. The circularity between *understanding, listening,* and *dialogue* shows the necessary dynamism and interconnection of the three methodological *pillars* contained in the case study. The tools and methods shown in the figure are therefore used for a double purpose: to augment the degree of perception and knowledge on the part of the population and various local actors; and to activate and/or strengthen interaction and communication between them.

The initial phase of activity therefore aimed to build a framework in citizens' perceptions of the urban impacts of CC and the broader regional consequences of local development.

That phase involved a dual approach: on the one hand, traditional and administrative representatives were involved as the largest holders of technical knowledge; on the other, community groups and residents of flood-prone areas were also engaged in the process.

The objective of this first phase was to facilitate involvement of the mayor, technicians, and traditional leaders in the spatial survey of the areas most exposed to environmental risks. After a preliminary literature review and mapping activity, field surveys and technical inquiries were used to establish a synoptic framework of the situation of vulnerable areas, using graphic elaborations to facilitate comprehension for the inexperienced.

Fig. 8.3 Photo of questionnaire administration

The results of this phase were an important bridge to subsequent activities involving interactions among local actors: graphic representations were a fundamental basis for discussion, and debate fed the collective production of maps of administrative distribution, land use, and the impacts of flooding.

The second phase focused on building trust between the people involved in order to create a favorable environment for exchange and dialogue. Meetings were organized that included a variety of activities and organizational approaches, such as panel discussions, semi-structured interviews, questionnaires, and collective mapping. Local people often assumed the role of moderator in order to stimulate dialogue among citizens and to facilitate empowerment processes.

The essential aim of this phase was to bring out the different sensitivities to climate change issues, thus facilitating awareness of spatial and environmental complexity.

A moment of great value for participants, both in terms of their accountability in the involvement process and their role in the management of the substantive issues in question, was the administration of the questionnaire to their fellow citizens.

The main objective of the questionnaire was to create a cognitive framework of overall social perception of territorial issues and, in particular, the consequences of floods.

Fig. 8.4 Photo of collective meeting

Members of the local youth association CREM (*Centre de Ressources Educationnelles Malika*) collaboratively developed the format and questions. Some of them were directly involved in administering the questionnaire to sample households in a 700-person neighborhood vulnerable to floods. The sample consisted of 30 respondents from 19 to 78 years of age and of both genders. The selection parameters for respondent households included number of members, number of workers within the household, gender, age, ethnicity, and average level of education. The questionnaire was carried out largely in the Wolof language, which is the most widespread in Senegal, especially in marginal areas where the level of education is very low (Fig. 8.3).

The questionnaire consisted of four parts that targeted specific and complementary factors: the first included questions about interviewees' personal information, such as marital status, employment, and housing, which was designed to compensate for the lack of data on the neighborhood of Medinatoul Salam; the second part addressed public services, and inquired as to respondents' access to water, electricity, and public facilities; the third explored the local economic dynamics of the formal and informal sector; and the last section specifically explored flooding problems in the district, and sought to bring out perceptions of the problem and the perspectives of each respondent.

Questionnaire data were elaborated with the collaboration of members of the CREM.

This phase of direct contact among inhabitants of different housing and social conditions stimulated dialogues and comparisons of various points of view, and it also increased interest level on environmental issues.

The knowledge sharing process continued through the structuring of meetings that would allow for dialogue between the various actors involved (Fig. 8.4). Meetings involved inhabitants of similar civic and social affinities in order to create a level of homogeneity that would facilitate spontaneous comparison of multifaceted perceptions. Participants' discussions were guided towards reasoning and exploration of the selected issues.

In the following interactive session, thematic group meetings were organized on various local issues that had been identified in previous stages.

In this case, the main driver was the interaction of heterogeneous levels of perception, knowledge, and, above all, liabilities relating to adaptation strategies and risk management for climate change. The main challenge was to critically, collectively, and jointly question potential adaptation measures and coping practices, while also maintaining a synoptic vision of spatial and environmental issues in local planning.

8.4 Main Results

The heterogeneity of the actors involved in the process and the limitations imposed by poverty, illiteracy, and the lack of information infrastructure—to name just a few factors—make the interchange and sharing of knowledge a real challenge.

The different levels of experiential education and training are clearly reflected in the cognitive and practical approaches of the people involved, as is reflected by the results of the interviews, questionnaires, and participatory activities.

Before discussing the results of the questionnaire on perception of CC, an overview of the social cognitive level regarding territorial and regional particulars is useful.

According to findings, people who do not work say they have no reason to move from the neighborhood where they live. Most inhabitants have a precarious work situation and they try to subsist through survival activities within the district. For this reason, many residents have not moved from their own neighborhood and a lot of them are not familiar with other areas of the city.

In addition to triggering complex social and cultural dynamics, this cognitive bias creates an effective barrier to the spatial perception within the neighborhood of the rest of Malika. It also affects the level of awareness of other flood-prone areas in the municipality. This common factor could instead offer an important opportunity for empowerment of environmental victims.

For example, the lack of spatial awareness is evinced with respect to the administrative division of the municipality. The results of cognitive investigations highlight that most of the people do not know the actual district subdivision:

Fig. 8.5 Solutions applied to the perimeter of flooded areas

variability of citizens' information oscillates from eleven to fourteen quarters, while the administrative documents indicate twenty-six.

This factor clearly reflects the gap between official sources and the knowledge that is widespread among the population, and highlights the considerable inefficiency of the administration's communication and information channels.

The results of the questionnaire analyses with respect to the social perception of CC indicate that cognitive limits are still very severe.

The issues raised were assessed according to three main levels of interpretation: cognitive apparatus on the spatial impact of floods; evolution of the phenomenon over time; and consequent health effects.

In terms of awareness of the spatial effect, responses ranged from an attitude of indifference to natural disasters, described as the *NIMBY* position (Not In My BackYard), to a fairly realistic awareness of conditions of vulnerability. Most people do not believe that global and local human actions contribute to environmental disasters over time. The low ability to inscribe environmental problems in a broader framework of global change and in the specific context of the location of at-risk settlements is very high.

People's perception is that the flood event is essentially limited to its direct implications. They are mostly residents in flood areas who engage in simple control solutions, which in most cases involve intuitive techniques to stem the flooded surface with recycled materials, partly also to prevent children from accessing it.

According to the questionnaire results, both collective and individual actions are taken. The former consist in constructing small barricades in order to block, reduce, or divert rainwater runoff (Fig. 8.5). In these situations, those who live close to major flooded areas support the initiative. In public spaces, people also organize lines of tires on the ground on which to walk in order access flooded zones.

Other solutions are applied at the household level. Where possible, families change or provisionally close the openings in outer walls to limit water-logging, and raise the level of the flooring both outside and inside the house.

Most people are fairly conscious of the serious health consequences of stagnant water. There is now widespread awareness that the permanent water basin facilitates the proliferation of mosquitoes and human parasites and promotes the spread of disease to animals. There were no local practices in this area to contain such problems.

Certainly, aggravation of the flood phenomenon since the first event in 2005 and the progressive increase of water surface area prompt greater attention on the part of inhabitants. The elderly are particularly sensitive to the evolution of floods over time. Closely following the changes over the years, they show more solidarity with fellow victims of the phenomenon.

In general, results indicate that the perception of CC is highly related to factors such as age, experience, and training; young people seem to be less fearful of the consequences of floods than older inhabitants. Sometimes this is justified by the greater propensity of young people to move into other geographic areas. The resistance to abandon one's house is usually higher at the family level and is particularly common among women; this phenomenon extends to the regional level and is generally a significant obstacle to the implementation of national *recasement* programs.[2]

According to the interview results, the knowledge gap on environmental awareness also affects local administration. There appears to be a considerable lack of information and communications relating to the urban and social impacts of CC. Malika is a small-medium sized district within the Senegalese territory. One hypothetical explanation for this knowledge gap is inadequate access to communication channels and media, which are more commonplace in urban centers.

---

[2] Programs that involve the construction of new habitations and housing systems for effected populations.

In this framework, local authorities are becoming aware that local people's responses to environmental change are linked with the evolution of subsistence activities, and with external pressures on their habits and daily lives that prompt their desire to lead more "modern" lives. As outlined in the interviews with administrative representatives, in many cases people have developed strategies for coping with weather variations and floods. Example of such strategies include:

- Changes in plot locations and crop diversification in order to minimize the risk of harvest failure;
- Changes in food storage methods, such as drying or smoking, according to corresponding food availability;
- Transformations of living areas and mobility patterns to deal with climatic variability.

8.5 Observations and Challenges

The present chapter aims to explore the participatory approach as a decision making tool for planning Climate Change adaptation strategies that integrate social and cultural issues and perceptions of local impacts. The methodology and results of a participatory process, carried out in the pursuit of three main objectives, are presented below.

The first objective was (1) *to assess the degree of awareness and perception of local urban issues, in terms of both generalized and more specifically regional impacts of climate change.* This has been achieved through the implementation of social interaction activities. The complementary stages of the process facilitated gradual involvement of participants and local representatives in the construction of join activities and the achievement of shared results.

While the study's contributions in terms of increased knowledge are difficult to quantify, the process certainly has fostered an awareness of some organizational and technical aspects that had not previously been considered by the residents or the local administration.

Increased awareness of and information on environmental and sanitation issues, land use, and the particularities of informal settlements constitute an important development trend in the land management of Malika. Shared framework reconstruction opens the way for a gradual integration of knowledge and skills in adaptation decision-making strategies and, more generally, local development.

The second objective of this study is to provide recommendations on favorable methodology, or (2) *to demonstrate the application of a participatory approach to sharing knowledge as a potential tool in the decision-making processes of urban planning, based on case study of Malika.*

The review of participatory processes is positive, particularly if one considers that consultation and dialogue between the actors involved has stimulated a strong interest in addition to having aroused the desire to define shared measures to reduce settlements vulnerability to CC. Meetings were characterized by significant results that often overcame the limits which were sometimes created in the initial confrontation between people of different social standing. The inputs provided by each participant's involvement in exploring and discussing territorial management has sparked good adhesion to the activities.

The methodological approach can also be recalibrated for other cities in the metropolitan area of Dakar as well as other developing countries with similar environmental and socio-cultural contexts.

Lastly, the study sought to provide (3) *recommendations regarding the importance of collaborations among various urban actors when establishing local strategies and measures for adaptation to CC.*

To be effective, actions need a continuous process of information interchange and knowledge sharing among urban actors. Constant updating of data on climate change should be simultaneously carried out at different levels, from the scientific to the political context, as well as channels of communication with citizens.

Therefore, local governments, which are crucial in the decision-making strategies for CC adaptation, should follow several key guidelines: to ensure transparency in governance and communications related to the availability and variability of local resources; to provide a framework of individual and collective practices, which may be a range of shared assumptions; to act as mediators between external intervention and local experiences; and to ensure the effectiveness and fairness of the strategies to be pursued.

Under such circumstances, the participatory process promotes citizens' needs and directs local development in a way that facilitates the role of local authorities.

Further efforts will be needed to cost the impacts of climate change and to inform and sensitize local audiences. Given that most adaptation efforts will take place at the local and sub-national levels, social and institutional agents should experiment, imitate, communicate, learn from, and reflect upon the impacts of climate changes and adaptation pathways.

To strengthen the educational, informational, and communicative level, appropriate tools and methodologies must be identified that are calibrated to involve a variety of actors. The cultural dimension of local communities, for example, requires an interchange that is not based only on written texts. It also demands direct communication and oral transmission in accordance with traditional widespread practice. In order to facilitate this process, information must be specific to the site, user-friendly, and more inclusive of indigenous knowledge and current local practices.

At the same time, the wealth of indigenous knowledge must be accessible to experts, translated into different languages and forms so that it can be properly shared in scientific research and in the institutional world.

The platform or forum for the direct exchange of information, experiences, and best practices is a tool that is gaining success in international web communication. Nonetheless, digital barriers between partners in different countries can significantly limit sharing processes and reduce adhesion.

Thus, three complementary categories should support knowledge sharing on adaptation to CC. At the local level, knowledge enhancement should be considered an intrinsic right and a source of inputs into the formal science; institutions should promote the augmentation of scientific knowledge and strengthen interchange relationships, whose potential has been underestimated; and finally, as regards adaptation strategies, the possibility of *overscaled* local experiences should codify general principles.

8.6 Conclusions

Adaptation is a relatively recent and rapidly evolving field. The conventional approach has focused mainly on the production of information and knowledge guided by experts and scientists. In most cases—particularly at the community level—knowledge about adaptation grows out of practice and therefore requires special attention to the interaction between various individual and collective entities acting on a local scale.

Tackling the problem of climate change from the environmental position is undoubtedly a priority, but it is not the only action required. It must also be understood in connection with the social and cultural issues resulting from climate change, which in turn will necessitate increased efforts to identify tools and adaptation strategies.

The final challenge is therefore how to connect and deepen autonomous adaptation and planning in specific countries or regions with high levels of risk, and how to highlight social vulnerabilities in addition to potential physical impacts.

In this context, learning from local communities becomes fundamental, and working with them allows for definition of opportunities and the implementation of useful adaptation pathways.

References

ActionAid International (2006) Climate change, urban flooding and the rights of the urban poor in Africa: key findings from six African cities. ActionAid International, London

Adger WN, Lorenzoni I, O'Brien K (eds) (2009) Adapting to climate change: thresholds, values and governance. Cambridge University Press, Cambridge

Agrawal A, Perrin N (2008) Climate adaptation, local institutions and rural livelihoods. In: Adger WN, Lorenzoni I, O'Brien K (eds) Adapting to climate change: Thresholds, values, governance. Cambridge University Press, Cambridge, pp 350–367

Dinar A, Hassan R, Mendelsohn R, Benhin J (2008) Climate change and agriculture in Africa: impact assessment and adaptation strategies. Earthscan, London

Diop A (2004) Gouverner le Sénégal. Entre ajustement structurel et développement durable. Karthala, Paris

Direction de la Prévision et la Statistique (2004) Recensement général de la population et de l'habitat—décembre 2002—Résultats provisoires, in Projections de population du Sénégal issues du recensement 2002

Hellmuth ME, Moorhead A, Thomson MC, Williams J (2007) Gestion des risques climatiques en Afrique: Apprendre de la pratique. Institut international de recherche pour le climat et la société. IRI, Columbia University, New York

Hope KR (2009) Climate change and poverty in Africa. Int J Sustain Dev World Ecol 16(6):451–461

Maddison D (2006) The perception of and adaptation to climate change in Africa. CEEPA Discussion Paper No. 10

Ochieng BO (2009) Effective communication of science and climate change information to policy makers. IDRC, Nairobi, p 22

Pelling M, High C, Dearing J, Smith D (2008) Shadow spaces for social learning: a relational understanding of adaptive capacity to climate change within organisations. Environ Plann A 40(4):867–884

Piermay JL, Sarr C (2007) La ville sénégalaise. Une invention aux frontières du monde. Kathala, Paris

Satterthwaite D, Huq S, Pelling M, Reid H, Lankao PR (2007) Adapting to climate change in urban areas. The possibilities and constraints in low- and middle-income nations. Human settlements discussion paper series, theme: climate change and cities – 1, IIED

Simone A (2004) For the city yet to come. Changing African life in four cities. Duke University Press, Durham and London

Smit B, Burton I, Klein RJT, Wandel J (2000) An anatomy of adaptation to climate change and variability. Clim Change 45:223–251

Thompson M (2008) ICT and development studies: towards development 2.0. J Int Dev 20:821–835

UNFCCC (2007) Nairobi work programme on impacts, vulnerability and adaptation to climate change. UNFCCC, Bonn

Yuen B, Kumssa Y (2010) Climate change and sustainable urban development in Africa and Asia. Springer, New York

Part III
Urban Impacts of Extreme Weather Events. A Case Study: Maputo, Mozambique

Chapter 9
Climate Change Hazard Identification in the Maputo Area

Maurizio Bacci

Abstract As stated in the fourth report of the Intergovernmental Panel on Climate Change (IPCC), climate change is affecting temperatures, sea levels, and storm frequencies in the entire world. While changes in average conditions can have serious consequences by themselves, the main impacts of climate change will be felt through weather extremes and the consequent risk of natural disasters. This chapter provides an overall picture of the climate conditions in the Maputo region, through the analysis of climatic data from the Maputo-Mavalane station (1960–2006). The current climate dynamics are analyzed and future climate scenarios are briefly considered, based on the literature of Mozambique. The study is especially focused on the aspects that most influence the management of a large city like Maputo. As such, attention is centered on the analysis of intense phenomena. The aim of this work is to contribute to local administrators' understanding of climatic phenomena and their processes.

Keywords Climate change scenarios · Disasters risk · Scenario analysis · Adaptation planning · Maputo

9.1 Climate Characterization of Maputo

The climate of Mozambique is mostly tropical, characterized by two seasons: a cool and dry season from May to September, and a hot and humid season between October and April.

Maputo city is situated on the west side of Maputo Bay, protected by the island of Inhaca. The climatological data, collected by the Instituto Nacional de Meteorologia (INAM) and used in this study for the period 1960–2006, come from the

M. Bacci (✉)
National Research Council—Institute of Biometeorology (IBIMET), Via G. Caproni 8, 50145 Florence, Italy
e-mail: m.bacci@ibimet.cnr.it

S. Macchi and M. Tiepolo (eds.), *Climate Change Vulnerability in Southern African Cities*, Springer Climate, DOI: 10.1007/978-3-319-00672-7_9, © Springer International Publishing Switzerland 2014

Table 9.1 Average monthly temperature and precipitation, Maputo-Mavalane (1960–2006)

Month	Jan.	Feb.	Mar.	Apr.	May	Jun.	Jul.	Aug.	Sep.	Oct.	Nov.	Dec.
Av. Min temp (°C)	22.3	22.3	21.5	19.2	16.2	13.6	13.2	14.7	16.7	18.4	19.9	21.4
Av. Max temp (°C)	30.9	30.7	30.2	28.8	27.3	25.5	25.2	26.3	27.3	27.8	28.7	30.2
Rain (mm)	160.4	139.9	96.3	49.9	26.3	15.2	17.6	13.1	33.4	56.1	88.9	87.2
Av. n. of rainy days	11	10	10	7	4	3	3	3	5	8	11	11

Maputo-Mavalane station. The station is located at 25°55′ S and 32°34′ E and records air temperature (min and max) and daily precipitation.

The temperature distribution in Maputo is almost constant, as shown in Table 9.1. The average maximum temperature varies from the warmest month, January (30.9°C) to the coolest month, July (25.2°C), and the average minimum temperature varies from 22.3 °C (January) to 13.2 °C (July).

Maputo is a relatively dry city, averaging 799.5 mm of precipitation per year during 1971–2000, or 784.3 mm/year during the entire 1960–2006 period. Rain distribution is characterized by high variability, with a maximum of 1299 mm/year in 1966 and a minimum of 288 mm/year in 2003. The city has a relatively short rainy season, lasting from November to March, with the wettest month in January, as shown in Table 9.2. With respect to the agro-ecological Köppen-Geiger climate classification zone, Maputo is Equatorial with a dry winter (Kottek et al. 2006).

The present climate study focuses on identifying the most intense rainy events and calculating the indicators for the length of dry spells. For each year of the historical series, the indices show some characteristics about rainfall and temperature distribution. In particular, the rain indices are: (i) the total cumulative precipitation; (ii) the Simple Daily Intensity Index (SDII), calculated by dividing the amount of rainfall by the wet days; (iii) the maximum number of wet days; (iv) the maximum number of consecutive dry days; (v) the annual count of days in which daily rainfall is more than 10 mm and (vi) more than 20 mm; (vii) the total annual precipitation when the rainfall rate is in or above the 95th percentile and (viii) the 99th percentile of the 1961–1990 daily rainfall data; and (ix) the annual maximum 1-day and (x) consecutive 5-day accumulated precipitation.

For each index, the least squares linear trend was calculated in order to evaluate the magnitude of the trend. A nonparametric Kendall's tau for monotonic trends was calculated for the series of indices using statistical software (Wessa 2011). Kendall's tau measures the strength of the relationship between the two variables, making no assumptions about the distribution of the data or the linearity of any trends. Like other measures of correlation, Kendall's tau assumes a value between −1 and +1. A positive correlation signifies that the values of both variables increase together, or decrease together. On the other hand, a negative correlation signifies that as one variable increases, the other variable decreases.

Confidence intervals can be calculated and hypothesis testing carried out with the help of Kendall's tau. The main advantages of using this measurement are that the index distribution has better statistical properties, and Kendall's tau can be

Table 9.2 Average monthly distribution of rainy days, rainy days above 10 mm, and rainy days above 20 mm (1960–2006)

Month	Jan.	Feb.	Mar.	Apr.	May	Jun.	Jul.	Aug.	Sep.	Oct.	Nov.	Dec.
Rainy days	11.0	10.5	10.1	6.8	4.0	3.2	3.0	3.1	4.5	8.1	10.9	10.6
Rainy days >10 mm	3.5	3.0	2.5	1.3	0.8	0.5	0.4	0.3	0.9	1.7	2.5	2.3
Rainy days >20 mm	2.2	1.8	1.5	0.6	0.4	0.1	0.2	0.1	0.3	0.7	1.3	1.1

Table 9.3 Annual rain indices—Kendall's tau trend statistics (1960–2006)

	Kendall tau	2-sided p value	Score	Var (Score)	Denominator	Probability (%)
Yearly rain	−0.077	0.452	−83.000	11891.0	1081.0	54.8
No. of rainy days	−0.200	0.051	−214.000	11868.7	1071.5	94.9
Max Cons. wet	−0.059	0.587	−59.000	11428.3	996.7	41.3
Max cons. dry	0.176	0.084	189.000	11863.7	1071.0	91.6
Rain days >10 mm	−0.068	0.514	−72.000	11816.7	1055.2	48.6
Rain days >20 mm	−0.055	0.606	−57.000	11768.3	1044.4	39.4
Rain days >95 p	−0.079	0.441	−85.000	11891.0	1081.0	55.9
Rain days >99 p	0.011	0.920	12.000	11890.0	1080.5	8.0
Max daily rain	−0.110	0.279	−119.000	11889.0	1080.0	72.1
Max 5-days rain	−0.045	0.660	−49.000	11891.0	1081.0	34.0

interpreted directly in terms the probability of observing concordant and discordant pairs. Contrariwise, outliers, unequal variances, non-normality, and nonlinearity unduly influence the Pearson correlation.

The results of the rainfall indices analysis in Table 9.3 reflect a negative trend for the number of wet days per year (2 fewer days in 10 years), with a probability of 94.9 % over the historical series, and in a positive trend for the maximum number of consecutive dry days (2 more days in 10 years) with a probability of 91.6 %.

These results are perhaps to be expected. In fact, the trends are difficult to detect due to the large interannual and decadal-scale variability of precipitation over the African region (New et al. 2006).

With respect to climatology, cooling episodes are not significant for Maputo. The lowest temperature recorded in the analyzed period was 4.3 °C in 1963, and the minimum temperature dropped below 10 °C for six consecutive days only once, in 1964. On the other hand, rising temperatures and heat waves could represent a problem, as they can aggravate the health conditions of the population. The highest temperature recorded was 43.7 °C and the average annual temperature maximum is 40.7 °C.

Regarding temperature indices, the trend analysis in Table 9.4 shows a more robust magnitude in comparison with rainfall. For each index, trends and correlations have been calculated with Kendall's tau test.

A positive trend was detected for the number of days in which the temperature was above 30 °C and the number of nights during which the temperature was

Table 9.4 Annual temperature indices—Kendall's tau trend statistic (1960–2006)

	Kendall tau	2-sided p-value	Score	Var (Score)	Denominator	Probability (%)
Freq. days above 30°	0.26	0.01	278.00	11874.6	1073.4	98.9
Freq. nights above 20°	0.54	0.00	584.00	11879.3	1075.4	100.0
Cool nights freq.	−0.48	0.00	−519.00	11871.6	1071.9	100.0
Cool days freq.	−0.31	0.00	−337.00	11864.3	1069.9	99.8
Warm nights freq.	0.54	0.00	581.00	11875.6	1073.9	100.0
Warm days freq.	0.10	0.35	103.00	11863.0	1068.9	65.1

above 20 °C, over an entire year. At the same time, the study indicates a negative trend in the frequency of cool nights and cool days, which are defined respectively according to the percentage of days when then minimum and maximum temperatures were in the 10th percentile or lower during the 1961–1990 period.

The analysis also revealed a positive trend in the frequency of warm nights, which is defined as the percentage of days when the minimum temperature is in or above the 90th percentile for the 1961–1990 period, although warm days reflected a non-significant trend in the historical series.

Part of this trend may be explained by the urbanization of Maputo from 1960 to 2006. The urban heat island phenomenon has been documented for many cities with varying populations, topographies, and climate regimes (Bornstein 1968; Li et al. 2004; Giridharan et al. 2004). Many causes contribute to the urban heat island effect, particularly the replacement of green areas with buildings, which impacts the radiation balance (due to the heat capacity and conductivity of building and paving materials, the increased absorption of short wave radiation, anthropogenic heat sources, etc.) (Oke 1982). For this reason, it is very difficult to quantify the influence of urban effects on the climatic trends recorded in the Maputo-Mavalane meteorological station time series.

This study has also analyzed the links between the El Niño/La Niña-Southern Oscillation and the rainfall and temperature regime in Maputo, considering the possible effects of teleconnections between El Niño and African precipitation (Nicholson and Kim 1997). The Southern Oscillation is the atmospheric component of El Niño. It is an oscillation in surface air pressure between the tropical eastern and the western Pacific Ocean waters (Zebiak and Cane 1987). The Southern Oscillation Index (SOI) measures the strength of pressure differences between Tahiti and Darwin, Australia. El Niño episodes are associated with negative SOI values, which indicate that the pressure difference between Tahiti and Darwin is relatively small. The monthly SOI values of are available at the NOAA Climate Prediction Centre.[1]

A correlation analysis was conducted for the SOI and the temperature and rainfall values in Maputo on a monthly and an annual scale. The monthly comparison in the entire series (Fig. 9.1) shows a weak correlation between SOI and

---

[1] See http://www.cpc.ncep.noaa.gov/products/analysis_monitoring/ensostuff/ensoyears.shtml.

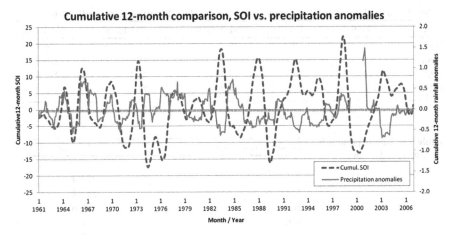

Fig. 9.1 Comparison between accumulated 12-month SOI and 12-month precipitation anomalies at Maputo meteorological station over the 1961–2006 period

temperature. Considering the time shift, the maximum correlation between the two indices is delayed by 4 months. Data indicates that positive (higher than average) SOI values detected in the Pacific have a slight positive effect on temperatures and rainfall in Maputo, with delays of 4 and 1 months, respectively. A negative anomaly in SOI (El Niño) has a positive effect on monthly precipitation in Maputo. The annual comparison also highlighted a weak negative correlation between SOI values and Maputo precipitation, with an R of −0.27.

9.2 Statistics of Extreme Events

9.2.1 Heavy Rains

As regards the most intense rainfall phenomena recorded at Maputo-Mavalane (Table 9.5), four episodes were characterized by an excess of 200 mm/day during the 1960–2006 period, and are sorted according to amount of rain in Table 9.5.

The year 2000 has been the wettest in the historical series, with 4 episodes above 100 mm/day and the single-day record of 336.8 mm. In 1966, 3 episodes of more than 100 mm/day were recorded, while in 1973 two such episodes were recorded. The other 14 episodes are distributed throughout the rest of the historical series.

The Gumbel distribution was calculated to characterize the distribution of extreme rainfalls at the Maputo-Mavalane station (Gumbel 1954). This methodology is used to model the distribution of extreme events (in this case, maximum rainfall) in a number of samples, and predict the chances those events will recur. Table 9.6 lists return periods for rainfall thresholds.

Considering the return period, the maximum rainfall episode of 2000 is expected with a frequency of less than once every 200 years (with an exact

Table 9.5 Most intense daily rain in the Maputo/Mavalane historical data (1960–2006)

Year	Month	Day	Max temp	Min temp	Daily rain
2000	Feb.	7	26.7	22.4	336.8
1966	Jan.	5	25	22.9	233.6
1967	Feb.	26	25.5	20.2	221.8
1978	Jan.	4	28.7	22.3	202.7

Table 9.6 Return periods for maximum daily rain

Return period (years)	Max daily rainfall
2	116.0
25	225.4
50	257.4
100	289.6
200	321.9
1000	397.2

estimated return period of 275 years). It is therefore reasonable to consider this episode a very rare case, assuming that climate conditions will remain the same in the future. Trends identified with respect to the most intense rain phenomena were thus deemed to be non-significant.

9.2.2 Drought

In order to characterize drought events in Maputo, the Standardized Precipitation Index, or SPI (McKee et al. 1993), shown in Fig. 9.2, was calculated for the historical series at 12-month intervals to individuate the drought phenomena that hit Maputo during the 1960–2006 period.

Using the SPI as an indicator, a functional and quantitative definition of drought can be established for the time scale chosen. A drought event for a 12-month time scale is defined here as a period in which the SPI is continuously negative and reaches a value of -1.0 or less. Drought begins when the SPI first falls below zero and ends when it regains a positive value. Drought intensity is arbitrarily defined for SPI values as follows: 0 to -0.99 is mild drought; -1 to -1.49 is moderate drought, -1.5 to -1.99 is severe drought, and -2.0 or lower is extreme drought. For Maputo, the SPI analysis detected five episodes of severe drought: 1964/65, 1970/71, 1982/83, 1991/92, and 2002/03 with an estimated return period for a severe drought phenomenon of once every 10 years.

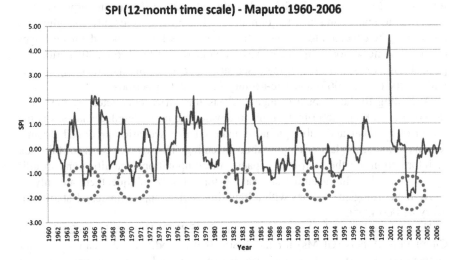

Fig. 9.2   Standardized precipitation index time series calculated for Maputo, 1960–2006, using a 12-month time scale

9.3 A Literature Review of Climate Change Scenarios for the Maputo Region

Mozambique is highly prone to climatic hazards and the government has long known that disaster prevention should be part of the development process. As a result, the government has included disaster risk reduction in all its main development policies and strategies. However, implementation still remains a challenge. The Ministry for the Coordination of Environmental Affairs (MICOA) produced the National Adaptation Program of Action (NAPA) in 2007. This critical policy document gives an overview of the situation in Mozambique and outlines the main measures to be undertaken to improve the country's resilience in the face of climate change. The NAPA, as well as all other official and non-official documents, considers three main climate hazards in Mozambique: droughts, floods, and tropical cyclones.

The Intergovernmental Panel on Climate Change (IPCC) is the leading international body for the assessment of climate change. The IPCC was established by the United Nations Environment Program (UNEP) and the World Meteorological Organization (WMO) to provide the world with a clear scientific view on the current state of knowledge on climate change and its potential environmental and socio-economic impacts. The main activity of the IPCC is to provide, at regular intervals, Assessment Reports on the state of knowledge on climate change. The latest is *Climate Change 2007,* the IPCC Fourth Assessment Report.

The main study conducted in Mozambique on the topic of climate change is the INGC Climate Change Report, coordinated by Mozambique's National Institute for Disaster Management (INGC) and published in 2009. The main purpose of this study was to understand how the climate of Mozambique is already changing and how it may be expected to change in the future. This INGC report details changes observed in the seasonal climate of Mozambique during the 1960–2005 period and presents downscaled future climate scenarios for Mozambique, focusing on the midcentury (2046–2065) and late-century (2080–2100) periods.

According to the climate change analysis for Africa conducted by the IPCC (Boko et al. 2007), the primary limit to developing more detailed analyses is the dearth of regional and sub-regional climate change scenarios produced through regional climate models or empirical downscaling. The main reasons for this lack of information are the restricted capacity of computational facilities and insufficient climate data. In the entire national territory of Mozambique, there are only 19 active meteorological stations with a long time series (approximately 1 station for every 42,000 km^2).

Under the medium–high emissions scenario (SRESA1B), used with 20 General Circulation Models (GCMs) for the 2080–2099 period, annual mean surface air temperature is expected to increase between 3 and 4 °C compared to the 1980–1999 period, with less warming in equatorial and coastal areas. When considering the temperature scenarios elaborated for three different temporal scales, one notes the convergence of rising worldwide temperature with a worst-case scenario for Mozambique of +4 °C in the 2080–2099 period. Despite the low accuracy of these outputs, it seems clear that a temperature increase is very likely.

Few studies involving climate change scenarios in Mozambique have been carried out. The most recent study was produced by Tadross in the INGC Climate Change Report (2009). One of the outputs of this study is the projected temperature change for 2046–2065, based on downscaled GCMs and shown in Fig. 9.3.

Both minimum and maximum temperatures are projected to increase in all seasons, as indicated in Fig. 9.3. It is therefore possible to predict that temperatures will rise between 1.5 and 3 °C by the 2046–2065 period.

Precipitation is characterized by a more complex distribution in the model outputs. In Mozambique, the IPCC AR4 models seem to predict a high confidence only for the southern part of the country, with a decrease of 0.1–0.2 mm/day in the 2080–2099 scenario.

For Mozambique specifically, the work of Tadross in the INGC—Climate Change Report (2009) offers a more detailed output for the 2045–2060 period (Fig. 9.4).

Model outputs indicate that rainfall can be expected to increase over most of Mozambique, during the DJF and MAM seasons, while less than approximate increases in evapotranspiration (0.1 mm/day) are predicted for the JJA and SON seasons. Higher increases in rainfall are projected in areas towards the coast, especially during the DJF season, with similar increases in coastal regions as well as towards Malawi during the MAM season. Greater uncertainty in the models is expected for the summer months (JJA), as it is more difficult to predict future rainfall patterns with confidence.

Fig. 9.3 Median changes in future maximum temperature from 7 GCMs (2046–2065 period); "+" and "−" indicate whether seasonal variability is expected to increase or decrease in the future (INGC 2009)

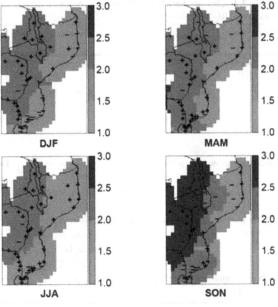

Fig. 9.4 Median changes in future rainfall (mm/day) from 7 GCMs; "+" and "−" indicate whether seasonal variability is expected to increase or decrease in the future (INGC 2009)

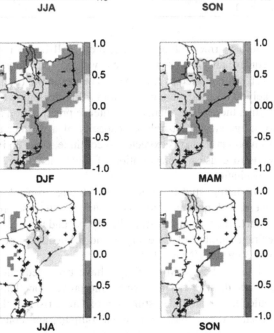

The probability of extreme events, including flooding and drought, is high for Mozambique and the Maputo municipality specifically. As such, disaster prevention is a high priority for the Maputo region, and subsequent studies will need to direct more attention to the future distribution of rainfall and high temperatures.

Table 9.7 Climate variables, likelihood of future trends and expected impacts (Short term horizon 2020–2030, long term horizon 2040–2060)

Climate trend	Projected likelihood	Impacts
Intensified rainfall	*Short term* Very likely	Increased flooding of buildings and homes
	Long term Likely	Increased flooding of roads
		Increased risk of flood-related causalities
		Increased risk of epidemics
		Beach pollution
Decreased annual rainfall	*Short term* Very likely	Decreased availability of water resources
	Long term Likely	Decreased stream flow
		Increased water demand
Higher maximum temperatures and heat waves	*Short term* *Long term* Likely Very likely	Increased illness in vulnerable populations
Increased wind speed and tropical storms	Likely	Infrastructure damage

In a warmer future climate, there will be an increased risk of more intense, frequent, and longer-lasting heat waves. The Maputo heat wave of March–April 2011, which lasted over a week, is one example of the extreme heat events that are likely to become more common in a warmer future climate. Along with the risk of drying, there is the chance that intense precipitation and flooding will increase due to the greater water-holding capacity of a warmer atmosphere. This has already been observed and is expected to continue, since an increase in world temperatures tends to cause a concentration of fewer, lengthier, and more intense periods of precipitation.

Modeling studies indicate that future tropical cyclones may become more severe, with greater wind speeds and more intense precipitation (Boko et al. 2007). In fact, such changes may already be underway, and statistics indicate that the average number of Category 4 and 5 hurricanes per year has increased over the past 30 years (INGC 2009).

Few studies of Mozambique have been dedicated to these critical aspects of climatic disaster risk awareness and prevention. This lack of detailed information could lead to an underestimation of future risks in this sector. For this reason, the present study is based on Global outputs, and considers their application to the municipal level as indicative of future trends.

In sum, the likelihood of future climatic trends and relative sets of expected impacts can be defined in order to determine appropriate risk reduction strategies. Table 9.7 provides climate variables and the expected impacts of each identified risk. The adaptive capacity of populations and the implementation of mitigation measures are not considered in this chapter.

9.4 Conclusion and Recommendations

Although climate change is a global problem, its impacts vary widely and are felt locally. This chapter has endeavored to provide useful information about the case of the city of Maputo and the types of challenges decision-makers will face in developing sustainable responses to various climate impacts. Historically, risk management strategies have relied on local experience, but with global climate change and rapid population increases, the future will not look like the past, and there is an urgent need to develop adaptive strategies. As this study illustrates, adaptation measures will be needed in Maputo to increase resilience to climate variability and to extreme weather events.

These findings demonstrate that increasing temperature and the decreasing number of rainy days is confirmed by Maputo meteorological station data and by GCM models. Meanwhile, changes in others climatic parameters are less statistically robust in sign and magnitude. For this reason, ongoing updating of new climate change scenarios will need to be produced by scientific community in order to define a clear vision of the future. Notwithstanding the difficulty of extrapolating the effects of the urban heat island based on regional temperature trends, given the concurrent expansion of the Maputo metropolitan area over the last 50 years, the heating phenomenon is expect to continue in the future, caused by a combination of urban heat island effects and broader climatic trends.

Despite the lack of detailed, local scale information on these issues, this study considers expected changes as a basis for planning future policies that prioritize interventions based on risk assessment. Spatial and temporal resolution is fundamental to climate change models. Best practice should use the same time period for planning and climate forecasts in order to adopt optimal choices in terms of magnitude of expected impact. As we move further in the future, the GCM introduces increasing degrees of uncertainty into the predictions: what is very likely in 2020 it is less likely in 2100. Especially for Africa, the GCM is not able to determine realistic initial conditions due to the weak measurement network in the region.

This study therefore proposes four key climate change risk identification procedures:

1. Regular review of the future climate scenarios presented in this study, particularly after the release of the Intergovernmental Panel on Climate Change's (IPCC) Fifth Assessment Report (scheduled for release in 2014).
2. Application of climate change scenarios as the basis for assessing risks in the initial stage of the risk assessment.
3. Use of appropriate temporal and spatial resolution of climate model outputs for correct planning. Where this is not possible, use of the closest scenario available and planning of policy mitigation measures accordingly.
4. Improve data acquisition networks and monitoring systems.

References

Boko M, Niang I, Nyong A, Vogel C, Githeko A, Medany M, Osman-Elasha B, Tabo R, Yanda P (2007) Africa. Climate change 2007: impacts, adaptation and vulnerability. In: Parry ML, Canziani OF, Palutikof JP, van der Linden PJ, Hanson CE (eds) Contribution of working group II to the fourth assessment report of the intergovernmental panel on climate change. Cambridge University Press, Cambridge, pp 433–467

Bornstein R D (1968) Observations of the Urban Heat Island effect in New York City. J Appl Meteor 7:575–582. http://dx.doi.org/10.1175/1520-0450(1968)007<0575:OOTUHI>2.0. CO;2

Giridharan R, Ganesan S, Lau SSY (2004) Daytime urban heat island effect in high-rise and high-density residential developments in Hong Kong. Energy Build 36(6):525–534. http://dx.doi.org/10.1016/j.enbuild.2003.12.016

Gumbel EJ (1954) Statistical theory of extreme values and some practical applications. Applied mathematics series 33. U.S. Department of Commerce, National Bureau of Standards

INGC (2009) INGC climate change report: study on the impact of climate change on disaster risk in mozambique. In: Asante K, Brundrit G, Epstein P, Fernandes A, Marques MR, Mavume A, Metzger M, Patt A, Queface A, Sanchez del Valle R, Tadross M, Brito R (eds) Main report. INGC, Mozambique

Kottek M, Grieser J, Beck C, Rudolf B, Rubel F (2006) World map of the Köppen-Geiger climate classification updated. Meteorol Z 15:259–263. doi:10.1127/0941-2948/2006/0130

Li Q, Zhang H, Liu XJ (2004) Urban heat island effect on annual mean temperature during the last 50 years in China Huang. Theor Appl Climatol 79(3–4):165–174. doi: 10.1007/s00704-004-0065-4

McKee TB, NJ Doesken and J Kliest (1993) The relationship of drought frequency and duration to time scales. In: Proceedings of the 8th conference of applied climatology, 17–22 Jan, Anaheim, CA. American Meterological Society, Boston, pp 179–184

New M et al (2006) Evidence of trends in daily climate extremes over Southern and West Africa. J Geophys Res 111:D14102. doi:10.1029/2005JD006289

Nicholson SE, Kim J (1997) The relationship of the El Nino-Southern oscillation to African rainfall. Int J Climatol 17(2):117–135

Oke TR (1982) The energetic basis of the urban heat island. QJR Meteorol Soc 108:1–24. 10. 1002/qj.49710845502

Wessa P (2011) Free statistics software, office for research development and education, version 1.1.23-r6. http://www.wessa.net/

Zebiak SE, Cane MA (1987) A model EI Nino-Southern Oscillation. Mon Weather Rev 115(10):2262–2278

Chapter 10
Impacts of Ocean Dynamics, Climate Change and Human Pressure on the East African Coast: The Case of Maputo

Carlo Brandini and Massimo Perna

Abstract Coastal erosion and loss of coastal environments are worldwide phenomena. These typical processes occur on different spatial and temporal scales, from river basins to coastlines, and from the ocean-atmosphere system to the global climate scale. All climate change scenarios foresee an increase in the global mean sea level in the next century, from a few tens of centimeters to over a meter. However, these scenarios are not sufficient to explain the accelerating erosion that already occurs today. In coastal areas, such change appears to be linked not only to sea level rise as a direct cause, but also to changing climatic conditions (changes in the rain distribution, winds, sea waves, etc.) and to increased human pressure on land (excessive use of weirs and dams along watercourses, loss of coastal dunes and areas of protective vegetation such as mangroves, etc.). The case of Maputo is quite informative, as none of the known effects of climate change is the main cause of the significant erosion processes that occur there today. Rather, this erosion is attributable to an altered balance between the contributions of sediment from neighboring river basins and to certain effects of coastal dynamics.

Keywords Sea level rise · Coastal erosion · Coastal flooding · Sediment balance · Maputo

C. Brandini (✉) · M. Perna
National Research Council, Institute of Biometeorology, and LaMMA Consortium for
Environmental Modelling and Monitoring Laboratory for Sustainable Development,
Via Madonna del Piano 10, 50019 Sesto Fiorentino, FI, Italy
e-mail: brandini@lamma.rete.toscana.it

M. Perna
e-mail: perna@lamma.rete.toscana.it

S. Macchi and M. Tiepolo (eds.), *Climate Change Vulnerability
in Southern African Cities*, Springer Climate, DOI: 10.1007/978-3-319-00672-7_10,
© Springer International Publishing Switzerland 2014

10.1 Introduction

Sea Level Rise (SLR) has been discussed extensively in recent decades. As early as the 1960s, the effects of SLR had already been noted by Bruun (1962) as a possible cause of the loss of coastline in many areas, especially along low shores. Over subsequent decades, increased attention was paid to SLR as a phenomenon closely linked to global warming. Many studies have aimed to determine its causes, and predict its future evolution (Jevrejeva et al. 2010; Rahmstorf 2007; Vermeer and Rahmstorf 2009).

Projected SLR scenarios are expected to have significant impacts on the economies of entire coastal regions, and may even affect the very survival of many settlements and major coastal cities. Such long-term scenarios require the design of new rules that will facilitate human adaptation to these changes and attenuate the level of risk, particularly since storm surges and floods in coastal areas will have greater impacts as SLR increases. To date, SLR has been widely recorded to be in the order of a few mm/year (2–4), although this rate seems to have increased in recent years. Looking at actual data, there is still no evidence of the sharp increase that is predicted by numerical models for the coming decades. However, this does not exclude the realization of forecasted scenarios, since SLR is far from being linear. Other processes acting to modify land-sea level changes include tectonics and eustasy, which should be carefully considered in a long-term perspective (Bruun 1988). Whatever the real SLR will be, it cannot be characterized simply as a greater elevation of sea level that will submerge coasts, but rather as a long-term process that will certainly have some degree of interaction with short-term processes. Today, in many parts of the world, we are already witnessing the loss of coastal environments due to processes that are globally referred to as *coastal erosion*.

One preliminary question arises directly from a rereading of Bruun's article: what is the actual role of the SLR observed in recent decades in terms of the loss of coastal areas (mostly sandy littorals) and more generally in what we call *coastal erosion*? The immediate answer is that, in some areas where spectacular coastal erosion phenomena are present, they have little to do with SLR, rather they are due to the alteration of an existing morphodynamic equilibrium. A coastline in morphodynamic equilibrium (i.e. not in erosion or in growth) is one where the sediment input and output are balanced. Such erosion is usually a local phenomenon, where reduced sediment input, and the continuation of causes that determine sediment transport (e.g. waves and coastal currents) act together. The input of sediments to a stretch of coast is due to sediment transport from an adjacent coast, or to the contribution of sediments derived from a watercourse. In the latter case, it is well known that many interventions made in the catchment area to inhibit the transport of river sediments have a negative impact on sandy shores. Along the African coast, this effect is particularly important because sediment transport depends on the river flow, and such flows have reduced considerably in recent years due to the documented reduction in rainfall, rising temperatures, and increased consumption of water for domestic, agricultural, and industrial use. Such increased consumption can also be considered an indirect effect of CC and the growing needs of the population. It is therefore *change*, in a broader sense, to which we refer to understand the complex phenomenon of coastal erosion.

This chapter refers to a case study of the Municipality of Maputo (Mozambique). To compare the two phenomena (coastal erosion and SLR) short- to medium-term dynamics and long-term trends are discussed. The former can be very significant and their influence on overall dynamics can cause intense, though possibly reversible, erosion. On a longer time scale, the changes indicated by SLR scenarios determine permanent and non-reversible effects. The first part of this chapter describes a model for interpreting the dynamics of Maputo Bay, based on dynamic variables (meteorological, hydrological, and oceanographic) and on their possible long-term changes.

It is subsequently argued that consideration of these phenomena must receive greater attention from planners, as knowledge of these dynamics is essential to successful planning of coastal protections and the reduction of vulnerability and risk for the population.

10.2 Description of Site

Maputo Bay is a subtropical embayment of the Indian Ocean, between 25° 50′ and 26° 20′ S, subject to large seasonal freshwater and tidal variations, and extends around 45 km from east to west and approximately the same from north to south. The bay is closed to the west by the Mozambique coast and to the east by the Machangulo peninsula and Inhaca Island. The bathymetry of Maputo Bay is shallow with an average depth of 5 m reaching ∼30 m at the ocean boundary. The bay opens to the shelf through an 18 km wide inlet (from the Macaneta sand dunes, the eastern boundary of the Incomati River, to *Ilha dos Portugueses* near the eastern side of the bay) (Fig. 10.1).

Maputo Bay is in itself a sort of natural harbor, well protected from wind and waves, particularly in the south, near the capital. The sea bottom is characterized by the presence of numerous sandbars and shallow areas, as well as by a large number of submarine channels. These channels in the bay's bathymetry were formed by the interaction between tides and the estuarine system, and in particular by the interaction between the Maputo River delta and the tidal inlet north of it (Mussa et al. 2003). The Machangulo peninsula and Inhaca Island form a barrier through a combined action of tides and aeolian deposition, thus protecting the river delta from the continental shelf.

Satellite images show plumes of turbidity at the main rivers' mouths, carrying in suspension mostly sandy and silty sediments. The shores in the area are either sandy beaches or mangroves at different stages of conservation.

Maputo Bay is subject to the impacts of a number of industries and several types of anthropogenic activities (heavy aluminum, salt farming, artisanal fishing,

Fig. 10 1 Maputo Bay and
its main riverine
contributions

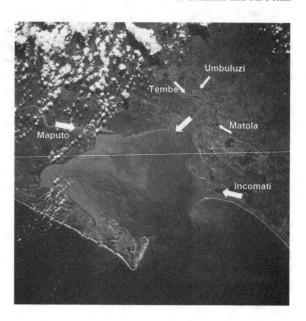

mangrove cutting for firewood) some of which are in conflict with each other in
terms of water use and sustainability (e.g. pollutants from domestic and industrial
wastewater disposal contrast with fishing activities).

10.3 Hydrographic Characteristics of Maputo Bay

From a hydro-morphological point of view, the bay is a very complex system,
whose dynamics need to be properly understood in order to predict future change.

Forcing by tidal action in Maputo Bay increases from neap (range ~ 0.5 m) to
spring tide (range ~ 3 m) with corresponding changes in the tidal currents
(from ~ 0.1 to ~ 1 ms^{-1}).

The region is also subject to strong seasonal rainfall: freshwater flow can be
very low in the dry season, and then peak at more than 10^3 m^{3}s^{-1} following
intense rains in the winter season. The volume of river discharge into the Indian
Ocean reflects, to a certain extent, the rainfall patterns in the region, thus rivers
draining high rainfall areas have relatively higher discharges. Consequently, the
region is characterized by the presence of large estuarine zones and extensive
mangrove forests.

Some of the most important river basins of the region conveying flow into
Maputo Bay are:

- Komati (Incomati) river, which enters the bay at its northern end;
- Maputo river, which enters in the South;

- Several smaller streams originating in the Lebombo Mountains, including the Matola (from the north), the Umbuluzi (from the west), and the Tembe (from the south), which meet towards the middle of the bay in an estuary generally known as the *English River* (formerly the *Espírito Santo River*). We will refer to this estuary simply as the Umbuluzi river.

The Incomati and Maputo rivers have large catchments, while the Umbuluzi's encompasses a relatively small area. Most of the flows are concentrated in the period from October to April, and have a pulse-like character.

The Incomati River has a total basin area of about 46,800 km^2, and an average yearly flow of 3.59 km^3 y^{-1}, 50 % of which is extracted mainly for agriculture. Concerning the flow regularization, the Incomati has the most significant dams, with 2060×10^6 m^3 of storage, followed by the Maputo and the Umbuluzi. The volume of water extracted from the basin has been rising. Historical data shows that, during the wet season, the mean flow in the Incomati was 80.69 m^3 s^{-1} with a recorded maximum 6827 m^3 s^{-1} during the catastrophic floods of 2000.

In comparison with the Incomati, the Maputo River has a basin area of about 29,800 km^2, has a lower peak flow, and is smoother throughout the wet season with less marked pulse discharges.

The rivers carry large quantities of freshwater and sediments to the bay. The main flows are diverted to the left due to the effect of the Coriolis force (quite evident from satellite imagery), so the flow exiting from the Incomati River tends to leave the bay, while that from Maputo is diverted along the western edge of the bay and tends towards north, joining its contribution with the Umbuluzi, in turn diverted to the north, and so feeding the sandy coast to the east of Maputo. Inside the bay, freshwater and sediments are stirred by the action of currents, and in particular by the interaction of oscillating tidal currents at the river entrance, and wind-induced currents (the so-called Ekman drift).

The action of surface waves is very weak. In fact:

1. The bay is closed with respect to almost all directions of the incoming waves, the fetch is very small in all directions, except for the northeast, which can be significant but hits only on the south coast of the bay (the mouth of the Maputo River);
2. A small percentage of waves from the Indian Ocean can enter the bay by diffraction, being much attenuated by this effect;
3. The bay is shallow, and this significantly contributes to reducing the height of the waves in the surf zone (where sediment transport is expected to increase).

However, while the shallows do limit the impact of wind waves on the coast, these do not prevent the entry of long waves (tidal surge) that can interact significantly with the hydrodynamics of the bay.

Satellite images show the presence of underlying forms that are shaped like large ripples, typical of areas with strong tidal excursions. The sediment balance is very delicate and liable to undergo changes as a result of any of the causes that promote the growth or erosion of the seabed and of the coastline. The same ecosystem has a strong dependence on freshwater input, not only because it is vulnerable to adverse effects from pollutants, but also because the sustainability of fishing activities (mainly shrimp) and the maintenance of the mangrove ecosystem are dependent on a minimum freshwater supply to the bay.

10.4 Hydrodynamic Modeling

The use of numerical models for understanding the oceanographic dynamics near the coast is one of the main tools for monitoring coastal seas, for planning measures to be carried out on the coast, and for developing future scenarios related to CC. It is worth noting that models are derived in large part from measured data, both remotely sensed and collected in situ. In coastal areas, oceanographic information that can be derived from satellite data has many limitations (e.g. for sea level). In situ data are crucial for this study. Unfortunately, apart from some data accessible through available publications, it is difficult to access local data, particularly updated bathymetric data at a good resolution, as well as hydrographical and water level data (to process information on tides, sea levels and morphodynamic parameters).

By contrast, a considerable amount of large-scale data is available from world atmospheric/ocean atlases and global numerical models, particularly as regards broad hydrographical features (mainly temperature and salinity), medium resolution bathymetry products, and atmospheric data regarding hydrodynamic forcing and tidal components. The local data needed for the present study was therefore derived through application of downscaling models to available large-scale data.

Maputo Bay is on the margin of the monsoonal regime, with large expected variations of atmospheric pressure and high winds. The bay is subject to considerable seasonal variations in freshwater input (~ 10–10^3 m^3 s^{-1}) and pronounced variations in tidal stirring power. During the dry season, the water column is fully mixed, with a weak horizontal density gradient and residual circulation mainly due to tidal currents. By contrast, during the wet season freshwater buoyancy was observed to induce marked horizontal salinity gradients and stratification, which is pronounced around the time of neap tides. Such a coastal system may be classified, as regards the freshwater input, as a monsoon-like regime (Lencart e Silva 2007): a sudden increase in rainfall induces large river runoff and creates the conditions for a change in circulation patterns.

In this context, the main exchange controls are not only tidal shear diffusion mechanisms, but also residual currents coming from:

- interaction between tides, coast, and bathymetry;
- density gradients (created by river runoff and surface heating); and
- large-scale effects, such as slope in the mean sea level (MSL) imposed by strong winds, atmospheric pressure, and variability in the off-shore current and eddy structures, among others.

A model of Maputo Bay at 1 km of resolution has been developed. The model, which integrates the ocean hydrodynamic (primitive) equations, is forced by wind, atmospheric pressure, the astronomical tide, temperature and salinity gradients, and main riverine input (in particular the three main contributions to Maputo Bay). The model is based on the Regional Ocean Modeling System (ROMS) code, a state of the art numerical circulation model (Shchepetkin and McWilliams 2005) that has been specially designed for accurate simulations of regional ocean systems. ROMS has been applied for the regional simulation of many different regions of the world ocean (Marchesiello et al. 2003).

To characterize the scales of detail and to obviate the problem of downscaling information from the global scale to the coastal scale of interest in this study, we chose to use a chain of nested models, starting from a large-scale climatological model at 1/4° of resolution and forced by climatological atmospheric data (COADS).

As for the oceanographic context, Maputo Bay is a sub-regional complex system situated between the Agulhas Current and the eddies system of the Mozambique Channel. The large-scale model includes the high-pressure system that dominates the region, and the intensification of the westward coastal current, typical of many of the western continental margins. In a classical view of the currents, the North Madagascar Current would flow south through the channel to form the Mozambique Current and, to the south, the Agulhas Current. Today, such a view has changed and the low variability scenario has been replaced by a configuration with greater spatial and temporal variability, in the form of a train of non-permanent anticyclonic eddies.

An intermediate (1/16°) resolution model has been spun to provide the boundary conditions of the high-resolution model of the bay (1/80° is about 1 km) (Fig. 10.2).

Maputo Bay is subject to a strong semidiurnal tide with a marked spring to neaps ratio (Canhanga and Dias 2005). Following the model, currents inside the bay range from 0.1 ms^{-1} in neap tide (minimum range of excursion) to about 1 ms^{-1} in spring tide (maximum range), as it is also confirmed by in situ data (Lencart e Silva 2007). Tidal analysis of the historical surface elevation established by Canhanga and Dias (2005) shows a minimum range of 0.2 m and a maximum of 3.8 m.

The model shows that temperature has an annual warming cycle, while on the other hand salinity has a small degree of variation: this is confirmed by observed data (Lencart e Silva 2007). Also, tide-induced residual currents show the formation of a non-permanent (although recurrent) eddy, just in front of the Incomati estuary, probably determined by the mutual interaction between tidal forcing, density front, and the coastline shape.

Tidal amplitudes vary over the year between 80 cm during neap tides and about 300 cm during spring tides. This is correctly reproduced by the model, as are the frequencies of tidal oscillation (Fig. 10.3).

In the INGC report (2009), the Highest Astronomical Tide (with a return period of 1 year) is estimated at 178 cm above MSL, which is consistent with measurements and model results. Moreover, an extreme sea level of about 270 cm is

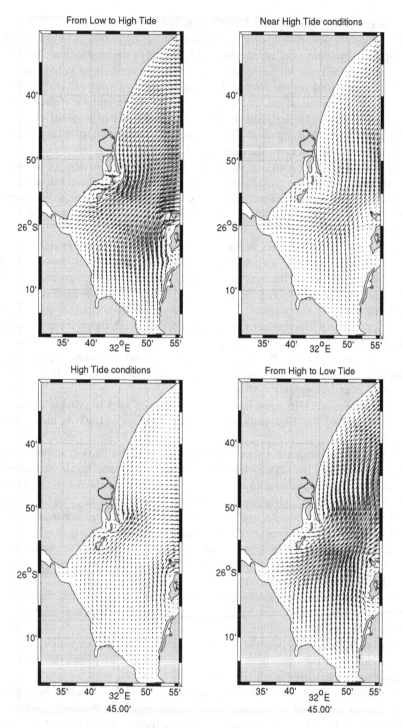

Fig. 10.2 Simulation of a Spring tide cycle in Maputo Bay, 16th November of a climatological year

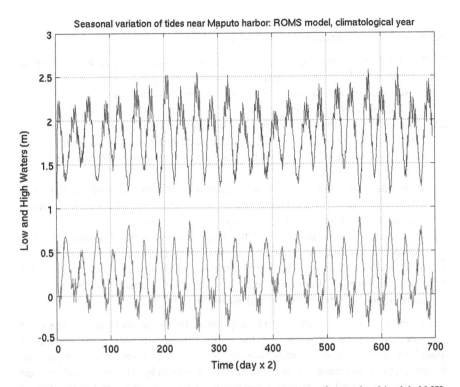

Fig. 10.3 Tidal variations calculated using the ROMS model. The reference level is global MSL

estimated for a return period of 100 years. These developments can reasonably be registered when a low-pressure system (e.g. a tropical cyclone) hits the coast and adds the effect of high tide to the inverse barometer effect, the latter due to strong winds causing a surge towards the coast.

10.5   Climate Change and Coastal Erosion

One of the biggest impacts that can be expected on the coast as a result of CC is brought on by the rising sea level. Although multiple data sources are now available to estimate the rise in average sea level, estimates of MSL based on in situ data do not support sufficiently reliable forecasts for the next century.

In particular, it is not possible to obtain data on sea level rise derived from historical hydrographical data over a long enough period to obtain valid extrapolations for the area in question. The only estimates of future SLR are from global models. Considerable efforts are being made by the scientific community to obtain more reliable estimates of future MSL. In coastal cities such as Maputo, this increase in MSL may have devastating consequences, especially if combined with other causes of rising sea level along the coast.

However, there is considerable variation among the projections being advanced at present in the international community:

(1) The IPCC estimates include an increase from 18 to 59 cm to be reached by 2100 (IPCC 2007). In comparison to the ongoing debate, these estimates are considered among the most conservative, and do not take into account a scenario of rapidly melting of ice.
(2) The melting of ice, as measured in multiple field studies, is associated with time scales in the order of several centuries and millennia. Most studies do not expect, for the next century, an increase of more than a few tens of cm to be contributed by melting ice.
(3) Many studies on sea level rise have been derived from paleoclimate studies, which have great validity, but are a poor basis for projecting scenarios of rapid changes over a few tens of years (Grinsted et al. 2009).
(4) Most of the studies that exceed the IPCC estimates of MSL rise by 2100 provide values that are usually lower or slightly higher than 1 m (Rahmstorf 2007; Rahmstorf 2010). However, some recent studies predict that sea level could rise by more than one meter this century if greenhouse gas emissions continue to escalate (Vermeer and Rahmstorf 2009).

This estimate is obviously not to be applied at the global scale, and differs from point to point along the coasts of the earth, interacting with local geological conditions such as tectonic uplift/subsidence, isostasy, or sediment compaction. Nevertheless, a maximum projected MSL rise of 1.25 m by 2100 has been deemed sufficiently conservative for the purposes of the present study, and that estimate has been applied to all subsequent determinations in this work.

Figure 10.4 depicts four scenarios, assuming an average sea level rise of 1.25 m in 2100:

1. the current situation;
2. the future scenario, in the case of an SLR of 1.25 m;
3. the current situation if the coast were be flooded by a storm surge during spring high tide conditions and with a return period of 100 years; and
4. the future scenario, in which MSL rise is compounded by a storm surge with return period of 100 years.

We can see that the majority of the city of Maputo would not be affected by the circumstances described above, with the exception of several newly urbanized areas to the east of Maputo (Costa do Sol). That area would not be particularly affected by rising sea level in itself, but it would be affected, even now, if a surge of great magnitude were to impact the coast, and that effect would be amplified in the future when combined with SLR.

The INGC report (2009) on the impact of climate change in Mozambique contemplates a much more severe scenario involving a rise in sea level equal to 5 m by 2100. It should be noted that none of the evidence presently available to the scientific community supports this hypothesis. Even more questionable statements appear in the same report to justify the need for coastal defenses against SLR.

Fig. 10.4 From *top* to *bottom*: current situation; mean sea level rise of 1.25 m in 2100; current situation with a surge return period 100 years; and future surge. Area of the port of Maputo on the *left*, Costa do Sol on the *right*

However, the only way to reduce the impact of sea level rise on human activities is through the rules of good building and good urban planning, and not through coastal defenses. The municipality should simply avoid the building of homes and neighborhoods in areas that are less than 4–5 m above MSL.

As in many other world areas, coastal erosion in the Maputo Bay is due to the combination of man-made and natural processes. Among the most significant anthropogenic activities are the damming of rivers (which causes a sediment deficit in the inputs to the coastal zone) and overexploitation of the littoral areas. Other conditions being equal (i.e. ignoring the causes attributable to human behavior), natural causes are likely to determine the main effects of CC. These natural processes include floods, storms, sea waves, and sediment transport (in both alongshore and cross-shore directions). Although the completion of this study would require more detailed data than those available, the model highlights major flows of water and sediment across the bay.

The concentration of suspended sediments within the bay was modeled using a number of Lagrangian particles released in the vicinity of the river mouths, and assuming a concentration of suspended sediments proportional to the input. The impact that a 20 % reduction of transport (freshwater and sediments) would have on the concentration of suspended sediment near the beaches was then estimated, taking as reference points the beaches near Costa do Sol. In practice, most of the sediment that nourishes the beaches to the east of Maputo seems to come from the mouth of the Umbuluzi river (10 %—i.e. half—of the total 20 % impact) and to a lesser extent from the Incomati and Maputo rivers. This highlights, once again, that the morphodynamic equilibrium within the bay is rather delicate and extremely sensitive to the reduction of sediments. In particular, the impact on sediment caused by the construction of river dams (such as the recently built *Pequenos Libombos* dam along the Umbuluzi river, close to Maputo) appears to be very high.

This low intake of fresh water flow (and sediment) at the river mouths, on the other hand, is well documented (UNEP 2009).

It should be noted that the coastal structures (groynes) put in place to protect the shoreline have been shown to be quite effective in protecting the coastal area: three such groynes, built on the eastern coast of Maputo, show an accumulation to the south (which seems to capture part of the sediments moved from the south, thus mostly coming from the Umbuluzi), while further north a slight accumulation has occurred on both sides, more pronounced on the northern side. This means that there is no clear trend in the direction of longshore sediment transport, with tidal oscillations moving comparable volumes of sediment in the two directions. The significant setback that the beaches have suffered in many places is therefore not due to alterations in the hydrodynamic regime, but rather to a reduced amount of sediments in the bay.

Coastal protection constructions (coastal walls) are effective in areas where the sea has already eroded much of the coast and the central problem is to protect existing infrastructure (e.g. along the Avenida Marginal). It should be noted that the conservation status of some natural elements along the coast, such as sand dunes and mangrove forests (often in combination) near the capital, is substandard. Recently,

Fig. 10.5 Examples of coastal erosion along the Avenida Marginal

many construction projects have been carried out in low areas occupied by mangroves. The restoration of these *natural* elements today is expensive or even impossible, However, in areas where mangroves are still intact, every effort should be made to maintain the natural elements that act as partial protection against erosion.

10.6 Conclusions

The projected long-term impact that climate change may have on the coast of Maputo, and other parts of the African coast, entails critical aspects that are insufficiently appreciated. Such aspects must, of course, be considered on a case-by-case basis. In the case of Maputo, the bay is part of a complex coastal system with a very delicate morphodynamic balance, determined by the joint action of tidal forcing and freshwater and sediment inputs from neighboring river basins (Maputo, Incomati, and Umbuluzi).

The reduction of sediment transport from hydrographical basins, caused by the intensive use of water (dams) and by the decrease in cumulative rainfall, is the most likely cause of coastal erosion, which is occurring mainly along the beaches east of Maputo. In the future, we can expect water consumption from rivers to increase, due to the economic development of this part of Mozambique and the consequences of global warming. This will undoubtedly lead to further reduction of the sediment transport, with serious consequences for the morphodynamic equilibrium of the beaches and of the bay in general.

The rise in sea level is currently a very controversial topic. Considered in isolation, the impact of this rise is not expected to be particularly serious for the city of Maputo, however important consequences are expected to occur along the beaches of the east coast (Costa do Sol), where intense urbanization is ongoing.

It is important to stress that there are no coastal defenses to counter the sea level rise. As such, a number of actions should be taken to avoid building in areas lower than 4–5 m above the actual MSL. Urban planning and urban rules are the only reliable tools with which to counter sea flooding phenomena, a risk that already exists today (e.g. in the event of a storm surge during a phase with very high tides).

Finally, it is worth noting that many elements of natural protection against coastal erosion (dunes, mangroves) in the Maputo region are now in poor condition, compromised, or destroyed (Fig. 10.5). There is therefore a need to avoid the over-exploitation and destruction of coastal sand dunes and mangrove forests, where they still exist. These environments are also important from an ecosystem standpoint as elements of outstanding environmental quality.

References

Bruun P (1962) Sea-level rise as a cause of shore erosion. J Water Harbor Div Proc ASCE 88(WW1):117–130

Bruun P (1988) The Bruun rule of erosion by sea-level rise: a discussion on large-scale two-and three-dimensional usages. J Coastal Res 4(4):627–648

Canhanga S, Dias JM (2005) Tidal characteristics of Maputo Bay, Mozambique. J Mar Syst 58:83–97. doi:10.1016/j.jmarsys.2005.08.001

Grinsted A, Moore JC, Jefrejeva S (2009) Reconstructing sea level from paleo and projected temperatures 200 to 2100 ad. Clim Dyn 34:461–472

Horton R, Herweijer C, Rosenzweig C, Lu J, Gornitz V, Ruane AC (2008) Sea level rise projections for current generation CGCMs based on the semi-empirical method. Geophys Res Lett 35:L02715

IPCC (2007) Climate change 2007: the physical science basis. In: Solomon S, Qin D, Manning M, Chen Z, Marquis M, Averyt KB, Tignor M, Miller HL (eds) Contribution of working group I to the fourth assessment report of the intergovernmental panel on climate change, Cambridge University Press, Cambridge, United Kingdom and New York, NY, USA, 996 pp

INGC (2009) Main report: INGC climate change report: study on the impact of climate change on disaster risk in Mozambique. In: Asante K, Brundrit G, Epstein P, Fernandes A, Marques MR, Mavume A, Metzger M, Patt A, Queface A, Sanchez del Valle R, Tadross M, Brito R (eds) INGC, Mozambique

Jevrejeva S, Moore JC, Grinsted A (2010) How will sea level respond to changes in natural and anthropogenic forcings by 2100? Geophys Res Lett 37:L07703. doi:10.1029/2010GL042947

Lencart e Silva JD (2007) Controls On exchange in a subtropical tidal embayment, Maputo Bay, Ph.D. thesis, University of Wales, Bangor

Marchesiello P, McWilliams JC, Shchepetkin A (2003) Equilibrium structure and dynamics of the California current system. J Phys Oceanogr 33:753–778

Mussa A, Mugabe JA, Cuamba FM, Haldorsen S (2003) Late weichselian to holocene evolution of the Maputo bay, Mozambique. XVI International union for quaternary research, Nevada, USA, July 23-30

Rahmstorf S (2007) A semi-empirical approach to projecting future sea-level rise. Science 315:368–370

Rahmstorf S (2010) A new view on sea level rise. Nature reports climate change. doi:10.1038/climate.2010.29

Sete C, Ruby J, Dove V (2002) Seasonal variation of tides, currents, salinity and temperature along the coast of Mozambique. Report. Unesco IOC Odinafrica and Centro Nacional de Dado Oceanograficos

Shchepetkin SP, McWilliams JC (2005) The regional ocean modelling system: a split-explicit, free-surface, topography-following-coordinate oceanic model. Ocean Model 9:347–404

UNEP/Nairobi Convention Secretariat and WIOMSA (2009) An assessment of hydrological and land use characteristics affecting river-coast interactions in the Western Indian Ocean region. UNEP, Nairobi Kenya 109p

Vermeer M, Rahmstorf S (2009) Global sea level linked to global temperature. Proc Nat Acad Sci USA 106:21527–21532

Chapter 11
Flood-Prone Areas Due to Heavy Rains and Sea Level Rise in the Municipality of Maputo

Sarah Braccio

Abstract The identification of flood-prone urban areas is of special interest to the majority of Sub-Saharan cities hit repeatedly by this type of disaster. This chapter presents the sources of information and a snapshot of methods used to identify areas prone to flooding due to heavy rains and sea level rise, information that is of use to the municipality of Maputo, Mozambique. This particular case is significant, given the disastrous events that have occurred there in the past few years and which can be attributed to climate change (INGC 2009). The sources of information available were compared and validated. The Modified Normalized Difference Water Index (MNDWI) calculated on the first available LANDSAT 7 ETM satellite images after the extreme rains of 6–8 February 2000 has proven to be the best method for the Maputo case. The snapshot method used allowed us to identify 57 km^2 of area exposed to flooding (16 % of the municipality). This result could be improved by georeferencing the flooded blocks systematically detected by district officers after each episode of heavy rain, and examining their correlation with the intensity of the physical events that led to their flooding.

Keywords Flood-prone areas · Landsat 7 ETM · Digital elevation model · Modified normalized difference water index · Maputo

11.1 Introduction

The areas identified as most exposed to heavy rains and sea level rise by Bacci, Brandini, and Perna (see Chaps. 9 and 10) must be pinpointed in order to use them in Ponte's (see Chap. 12) risk equation:

S. Braccio (✉)
Interuniversity Department of Regional and Urban Studies and Planning,
Politecnico di Torino, Viale Mattioli 39, 10125 Turin, Italy
e-mail: sarah.braccio@polito.it

S. Macchi and M. Tiepolo (eds.), *Climate Change Vulnerability in Southern African Cities*, Springer Climate, DOI: 10.1007/978-3-319-00672-7_11, © Springer International Publishing Switzerland 2014

$$R = (H * V * E)/A$$

$$R(\text{risk}) = [H\,(\text{hazard}) \times V\,(\text{vulnerability}) \times E\,(\text{exposure})]/A\,(\text{adaptation})$$

The present analyses aim to supply information that will help local authorities identify climate change adaptation measures (see Chap. 13).

Flood-prone areas can be identified using a hydrographic model. However, in Africa, large cities rarely have access to information on the nature of the soil, orography, daily precipitation in hours, or the consistency and state of maintenance of stormwater drainage, information necessary to create a functioning model. In the absence of such data, two different rapid methods can be used.

The first is the identification of those areas which, due to the terrain, tend to accumulate rainwater (see Degiorgis et al. 2012; Manfreda et al. 2011; Ho et al. 2010; Taubenbock et al. 2011).

The second consists in ascertaining, through on-site inspections, the areas flooded following heavy rains and where, if possible, the level of intensity is known.

The use of snapshot methods should not be considered an alternative to normal simulations using hydrological/hydraulic models, but they can be quick and useful tools for preliminary identification of areas subject to flooding, should the long and costly process of constructing a model is not feasible.

This chapter will consider two factors that are useful for calculating the risk of flooding: firstly, the basins where rainwater tends to stagnate and the areas subject to flooding following extreme tides in the municipality of Maputo (347 km^2); secondly, the pressure on flood-prone areas caused by runoff water originating from the secondary watershed to which each of those areas belongs.

Risk calculation drew on six sources of information available on four city districts of Maputo and their 54 *bairros*, excluding the less populated districts of Katembe and Inhaca.

The first was the digital elevation model (DEM) based on ASTER (GDEM) satellite surveys with 30 m resolution. The GDEM was obtained by stereoscopically comparing 1.3 million optical images from ASTER, which cover approximately 98 % of the Earth's surface. These images can be downloaded free of charge from NASA's EOS data archive or from Japan's Ground Data System. The DEM facilitated identification of the lowest areas within municipal borders.

The second source was LANDSAT TM images.

The third source was the catalogue of satellite images available on Google Earth.

The fourth source was the map produced by the *Instituto Nacional de Gestão de Calamidades* (INGC).

The fifth source was the list of areas flooded on 15 January 2011 as detected by the municipality's experts and the list of *quarteirão* (a minimal territorial entity, comprised of a few city blocks) flooded after the heavy rainfall of 15–20 January 2011 as detected by district experts and officers from the *bairro* (neighborhood).

The sixth source was the study produced by the USGS after the extreme rainfall (337 mm) of 7 February 2000 in order to identify flooded areas, though definite information on the methods used in that study was not available.

The final phase consisted in reconstructing the runoff pressure on identified areas, and comparing the exposed surface area to the surface area of the watershed to which it belongs. This is a simplification that presumes the same impervious surface rate, the same soil nature, and the same slope across the entire secondary watershed. Nevertheless, it is useful for calculating risk. The use of the satellite images and digital elevation models (DEM) freely available online and the simplifications adopted were aimed at developing a method that could be repeated notwithstanding limited financial means and could be managed by local authorities with no particular expertise (The World Bank Group AFTUW 2012).

Among the difficulties encountered was the lack of information on an urban scale. In an international context, Rainfall Detection Systems (Ajmar et al. 2011) use satellite images, like MODIS, to pick out bodies of water. However, those images have a spatial resolution that ranges from 250–1000 m and are therefore unsuitable for an urban context. According to the National Disaster Management Authority of the Government of India (2009), a scale of 1:10,000 and elevation information with an accuracy of 1 meter (RMSE) in rolling and hilly areas are considered appropriate when identifying flood-prone areas. Yet the digital land models available, such as the ASTER GDEM, reach a maximum accuracy of 10 m.

11.2 Areas Prone to Flooding Caused by Heavy Rainfall

The 1960–2006 daily rainfall series used by Bacci (see Chap. 9) shows 16 days with precipitation ranging from 100 to 200 mm and seven days with precipitation over 200 mm (Table 11.1). The former events recur, on average, every two years, while those over 200 mm occur approximately once every 10 years, and were thus defined as extreme.

The precise identification of areas prone to flooding due to extreme rainfall requires the preparation of a hydrological model that takes into account the conditions of water on the surface and underground. Such a model requires information such as the site's orography, the nature of the soil, its level of impermeability, the level of precipitation (especially during the first few minutes of rain), ground water levels, and the existing drainage system and its state of maintenance.

With respect to spatial aspects, three different types of hydrological model can be identified (Jha et al. 2011):

- one-dimensional models: these are simplified models in which only certain aspects are taken into account, such as water depth, velocity and runoff

Table 11.1 Maputo
1960–2006. Rainfall in mm
(Bacci, see Chap. 9)

Rainfall class (mm)	Days (No.)
1–49	3,950
50–74	70
75–99	32
100–199	16
Over 200	7

direction. Examples of this kind of model are beach profile models, HEC-RAS, LIS-FLOOD, and HYDROF;

- 2-D models, which consider runoff occurring concurrently to the main rainfall phenomenon. These models are used in topographically complex areas such as large floodplains, and they require high-quality data and long calculation times. TELEMAC 2D, SOBEK, DELFT 3D, and MIKE21 HD are examples of two-dimensional models;

- 3-D models, in which all three components of water velocity are taken into account. These are complex models that can only be used over small areas. Examples include FINEL 3D, FLUENT and PHOENIX.

As regards Maputo, apart from the DEM, the nature of the soil, and total daily precipitation, all other information is lacking.

Another method, one that would allow for quick identification of the main areas of rainwater stagnation, was therefore needed. Six types of information were used to develop such a method.

Firstly, a digital model of the terrain allowed researchers to identify the lowest areas within the municipality's borders (Fig. 11.1). This information was provided by the ASTER GDEM (ground resolution 30 m—1 pixel = 30 m). The Advanced Spaceborne Thermal Emission and Reflection Radiometer (ASTER) Global Digital Elevation Model (GDEM) covers the entire global land surface. Horizontal accuracy is approximately 30 m with 95 % confidence, and vertical accuracy is 20 m with 95 % confidence (Wang et al. 2011; Tarekegn et al. 2010).

The slope image of the Maputo area was provided by the DEM SRTM 90 m. The SRTM (Shuttle Radar Topography Mission) DEM has an absolute vertical accuracy of ±16 m and an absolute horizontal accuracy of ±16 m. Raster map layers of slope, aspect, and curvatures were generated as well as partial derivatives from a raster map layer of true elevation values (Fig. 11.1). Aspect was calculated counter clockwise from the east.

Second, researchers verified the possibility of identifying areas flooded by heavy rainfall on the basis of LANDSAT 4 multispectral images (with a maximum 30-m resolution). In particular, seven episodes of heavy rainfall between 1973 and 2011 were identified. In five of the seven cases, LANDSAT images were taken at least 19 days after the event, and up to 208 days after in the case of the disastrously heavy 225 mm rain of January 1978 (Table 11.2). Moreover, in three of the seven cases, cloud cover in these images was 30 % or more (Table 11.2).

Fig. 11.1 Maputo (*black border*). Orographic image from DEM SRTM 90 (formulated by Braccio 2011)

Table 11.2 Municipality of Maputo, Mozambique, 1973–2011

Rainfall Date	LANDSAT image mm	Image delay Date	L type	Cloud cover %	Quality
1973, Sep 28–30	319.9	1973, Oct 8	L 1–3 MSS	30	4
1976, Jan 39–31	313.3	1976, Feb 10	L 1–3 MSS	10	5
1978, Jan 3–4	225.1	1978, Aug 25	L 1–3 MSS	80	0
1984, Mar 23–25	219.7	1984, Jun 1	L 4–5 MSS	20	9
2000, Feb 6–8	551.2	2000, Mar 1	L 7 SLC-On	41	9
2000, Mar 8–11	214.2	2000, Apr 2	L 7 SLC-On	23	9
2011, Jan 12–15	200.0	2011, Feb 1	L 4–5 TM	7	0

LANDSAT images available in the archives and distributed by the USGS after seven episodes of heavy rain (>200 mm) (http://glovis.usgs.gov/; formulated by Braccio 2011)

The presence of cloud cover at the time when the LANDSAT images were recorded hindered calculation of the NDWI index, since the covered pixels appear as *no data* and the pixels overshadowed by the clouds have altered DN values. As such, the areas exposed to flooding could not be identified through satellite surveys, and researchers' only option was to deduce them from flooded areas following major events.

Fig. 11.2 Maputo,
Magoanine A-B-C *bairro*,
2000 (*left*) and 2010 (*right*).
Four months after the
disastrous rainfall of 6–8
February 2010, the area
outlined in black was still
flooded (*left*). Ten years later,
the basin which had been
previously flooded was found
to be completely built up
(*right, black outline*)
(formulated by Tiepolo 2011)

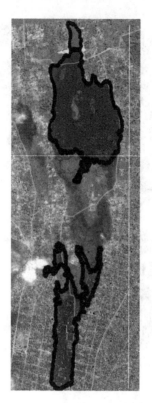

Though more costly, this analysis was possible through photo-interpretation of GeoEye-1 high-resolution satellite images, which, since 2000, allow for a multi-temporal analysis for the purpose of identifying flood-prone areas. The GeoEye-1 satellite has the highest resolution of any commercial imaging system and is able to collect images with a ground resolution of 0.41 m, re-sampling to 0.5 m (1 pixel = 0.5 m) in full color or the black-and-white mode. It collects multispectral or color imagery at 1.65-m resolution. The advantage of GeoEye-1 comes from orbit performance. Researchers verified the availability of free archive satellite images on Google Earth, which—though not available immediately after heavy rainfall—have the advantage of allowing the photographic interpretation of flooded areas. By contrast, LANDSAT images are available immediately following heavy rainfall, however they are subject to a cloud cover of 30 % or more, and as such cannot be used. The test carried out in the Magoanine A-B-C *bairro* after the disastrous rains of 6–8 February 2000 revealed the existence of a 1.6 × 0.8 km area that was still flooded 4 months after the heavy rains (Fig. 11 2).

Ten years later, the same area was completely parceled out and largely built up (Fig. 11.2 right). Over 800 building plots had already been completely developed.

Fig. 11.3 Maputo, 20 May 2000. LANDSAT 7 ETM multispectral satellite image taken 4 months after heavy rainfall, with the areas identified by the USGS (2000) study highlighted. According to the MNDWI, bodies of water are shown in grey and moist soil in white (formulated by Braccio)

Housing had been constructed in flood-prone areas without any drainage system to keep the site dry in case of heavy rainfall.

Third, following the photographic interpretation of available Google Earth images, a LANDSAT 7 ETM satellite image taken on 20 May 2000 using the Modified Normalized Difference Water Index (MNDWI) was also analyzed. That image shows flooded areas (Fig. 11.3 in grey) both in the Magoanine A-B-C *bairro* and in other parts of the city. The identification of moist soil and water according to the MNDWI index (Ho et al. 2010) can facilitate detection of flooded sites. This method allowed researchers to isolate flooded areas of LANDSAT images obtained during the flood season and water-saturated areas of images obtained during the rainy season with high potential for inundation.

Fourth, the data gathered through on-site inspections, carried out by municipal officers following heavy rains, was analyzed. This comprised two sources of information collected during the heavy rainfall of February 2011. The first source was a report entitled *Municipal Survey of Flooded Areas in February 2011* (Fig. 11.4), while the second was a map entitled *Bairros Frequentemente Inundados Devido a Chuvas Intensas* produced by the INGC-CENOE (Fig. 11.5).

Fig. 11.4 Maputo 2011.
Flooded areas and streets
after the flooding of 15
January 2011 (Maputo
Municipality 2011;
formulated by Ponte 2011)

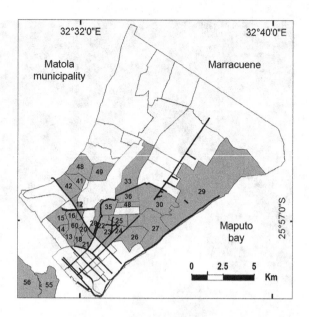

Fig. 11.5 Maputo 2010.
Frequently flooded *bairros*
(INGC CENOE 2010;
formulated by Ponte 2011)

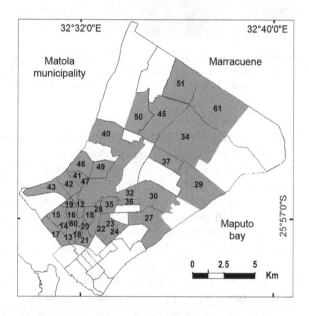

Both the Municipality's report and the INGC-CENOE map involved collection
of diverse types of information (concerning streets, particular points or entire
bairros). They report on 112 and 93 ha of flooded areas, respectively. In both cases,

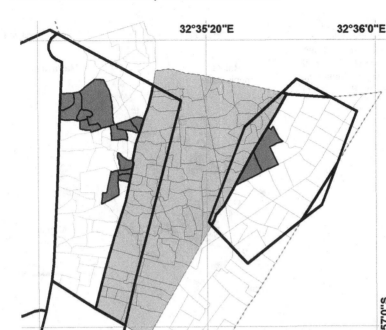

Fig. 11.6 Maputo 2011. Kamaxaquene municipal district. Flooded *quarteirões* according to the survey carried out following the heavy rains of 15–20 January, and flood-prone areas according to the USGS. The *grey bairro* has 15 unspecified flooded *quarteirões* out of 85. (Formulated by Ponte 2011)

the information provided was not particularly useful for an analytical identification of the areas at risk.

Fifth, the areas flooded following heavy rains were examined in a report drafted by each district. These analytical reports identify single flooded *quarteirão*. Unfortunately, this information is difficult to access for various reasons (one of which was the unfortunate theft of the computer where they were stored, another was the destruction of the archive). The present study analyzed information concerning the Kamaxaquene district and researchers were able to verify which *quarteirão* were flooded according to the district's inspection following the rainfall of 15–20 January 2011 (Fig. 11.6). The result was a surface area significantly smaller than the one highlighted on the basis of the district's orography, but nevertheless within the areas this study considers to be exposed to flooding.

Finally, this study took into account the flooded areas identified by USGS in 2000 (Figs. 11.7 and 11.8). Those areas constitute 8 % of the total surface area of the four districts of Maputo considered in the present study (which excludes the districts of Katembe and Inhaca) (Table 11.3).

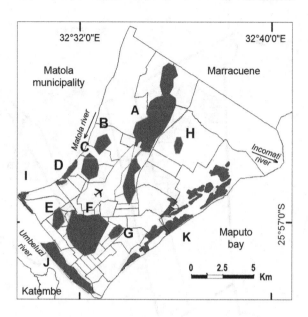

Fig. 11.7 Maputo 2000. Flood-prone areas due to heavy rains (*A–H*) and due to sea level rise (*I–K*) (USGS 2000; Brandini estimation; formulated by Ponte 2011)

There is considerable overlapping of the lower parts of the secondary watersheds and the areas identified by the USGS study.

11.3 Areas Prone to Flooding Due to Sea Level Rise

Brandini and Perna (see Chap. 10) have identified flood prone areas due to rising sea levels based on the hypothetical extreme tide of 270 cm with a recurrence rate of 100 years identified by the INGC. In addition to the extreme tide, a sea level rise of 100 cm is projected for 2100. Brandini and Perna (ibid.) identified the flood-prone area (Fig. 11.10 black border) with a rise of 370 cm of the sea level above the average tide level and where Ponte (see Chap. 12) identified built-up areas in September 2010 (Fig. 11.9 white border) (Table 11.4).

It is useful to keep in mind that the flood-prone areas identified in this way are quite different from those considered flood-prone in the Maputo Municipality Master Plan.

In total, flood-prone areas due to heavy rainfall and extreme tides cover approximately 16 % of the continental territory of the municipality of Maputo, excluding the districts of Katembe and Inhaca.

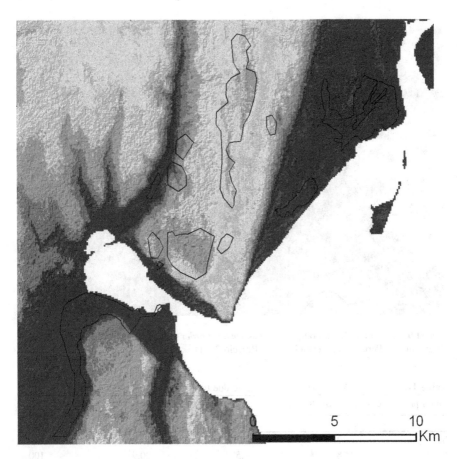

Fig. 11.8 Maputo 2000. Overlying DEM SRTM 90 m, and flood-prone areas as identified by the USGS in 2000 (formulated by Braccio 2011)

Table 11.3 Maputo. Characteristics of the eight areas prone to flooding following heavy rainfall

Flood prone areas due to heavy rains								
	A	B	C	D	E	F	G	H	$\sum$
ha	1,383	127	192	112	117	778	79	68	2,856
%	24.1	2.2	3.3	2	2	13.6	1.4	1.2	100
%	–	–	–	–	–	–	–	–	12.4

11.4 The Pressure of Secondary Watersheds on Flood Prone Areas

Flood-prone areas are such because of their position at the bottom of several of secondary watersheds. They receive water not just from rainfall but also from rainfall runoff on the surface of each secondary watershed. The larger the

Fig. 11.9 Maputo 2010. Flood-prone areas (*black border*) and built-up areas (*white border*) (Brandini and Perna 2011; formulated by Braccio 2011)

Table 11.4 Maputo. Areas exposed to flooding due to a sea level rise of 370 cm

Flood prone areas due to S-LR			
	I	J	K	Σ
ha	1,423	256	1,197	2,876
%	24.8	4.5	20.9	100
%	–	–	–	12.6

watershed is compared to the exposed surface area, the greater the pressure will be on the latter.

The level of runoff depends on the soil texture, the terrain's gradient, and the level of impermeability (due to construction, the paving of open spaces and roads, etc.). Although these factors can be measured, the information needed to do so was not available at the time of this study. In stead, for the purpose of risk analysis, the secondary watersheds belonging to each flood-prone area were identified. The exposed area was then compared to its secondary watershed, assuming a relative amount of uniformity in the factors determining the level of runoff in all secondary watersheds.

Such comparison indicated that secondary watersheds range from two to five times the surface area exposed to flooding (Fig. 11.10, Table 11.5).

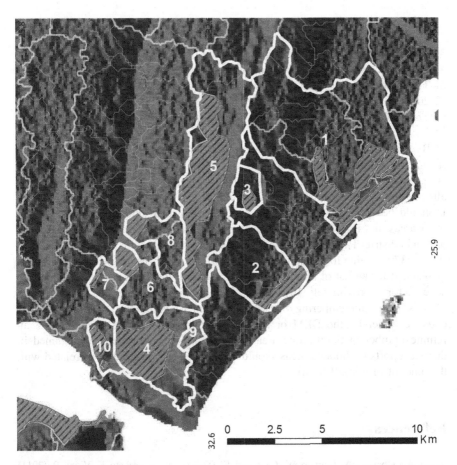

Fig. 11.10 Maputo 2000. Flood-prone areas (ha) and their watersheds (formulated by Braccio 2011)

Table 11.5 Maputo 2011. Size of flood prone areas due to heavy rain compared to their secondary watersheds

	Dimensions of the watershed							
	A	B	C	D	E	F	G	H
FPA (ha)	1,383	127	192	112	117	778	79	68
Watershed (ha)	2,702	96	736	277	304	1,364	116	253
W/FPA	2	4.7	3.8	2.5	2.6	1.7	1.4	3.7

11.5 Conclusions

This study has identified the areas exposed to flooding due to heavy rainfall (>200 mm) and extreme sea level rise (tide + sea level rise).

The methodology evaluated the six types of information available:

1. A DEM to identify the lower parts of each secondary watershed
2. LANDSAT TM images taken after heavy rainfall
3. The catalogue of satellite images available on Google Earth
4. The INGC map of flood-prone areas
5. The district reports on flooded *quarteirão* following the heavy rains of January 2011
6. The USGS study drafted after the heavy rains of February 2000.

By analyzing this information, researchers have deduced that the basins of the secondary watersheds identified using the DEM and confirmed by other sources are consistent and remain the best sources for identifying flood-prone areas. This allowed for identification of eight areas exposed to flooding due to heavy rainfall. It should be specified that these are areas where water tends to stagnate (unlike other areas, not included this study, where flash floods occur, such as the Central *bairro* in district 1). Other areas that will be subject to flooding should the sea level rise by 370 cm due to the combined effect of sea level rise and extreme tides are also added to the list of flood-prone areas. Overall, the areas exposed to flooding account for approximately 16 % of the municipality's surface area.

It is worth remembering that altimetric micro variations, which cannot be detected either by the DEM or the technical map available, can determine an infinite number of flood-prone areas. This is why the information contained in district reports on flooded areas should be mapped using GIS and correlated with the amount of rainfall (mm).

References

Ajmar A, Albanese A, Camaro W, Cristofori EI, Dalmasso S, Disabato F, Vigna R (2011) Extreme rainfall detection system for UN WFP in WSN12. In: 3rd WMO/WWRP international symposium on nowcasting and very short range forecasting, Rio de Janeiro, 2012

Degiorgis M, Gnecco G, Gorni S, Roth G, Sanguineti M, Taramasso AC (2012) Classifiers for the detection of flood-prone areas using remote sensed elevation data. Int J Appl Earth Obs Geoinf 470–471(2012):302–315

Ho LTK, Umitsu M, Yamaguchi Y (2010) Flood hazard mapping by satellite images and srtm dem in the Vu Gia–Thu Bon alluvial plain, Central Vietnam. In: International archives of the photogrammetry, remote sensing and spatial information science, vol XXXVIII, Part 8. Kyoto Japan

INGC (2009) Synthesis report. In: Van Logchem B, Brito R (eds) INGC Climate change report: study on the impact of climate change on disaster risk in Mozambique. INGC, Mozambique

Jha AK, Bloch R, Lamond J (2011) Cities and flooding. a guide to integrated flood risk management for the 21st century, The Word Bank, Washington

Manfreda S, Sole A, Di Leo M (2011) Detection of flood-prone areas using digital elevation models. ASCE J Hydrol Eng 16(10):781–790

National Disaster Management Authority Government of India (2009) High Resolution 1:10,000 scale Mapping Strategy of Multi-Disaster Prone areas in India Report

Tarekegn TH, Haile AT, Rientjes T, Reggiani P, Alkema D (2010) Assessment of an ASTER-generated DEM for 2D hydrodynamic flood modeling. Int J Appl Earth Obs Geoinf 12(2010):457–465

Taubenbock H, Wurm M, Netzband M, Zwenzner H, Roth A, Rahman A, Dech S (2011) Flood risks in urbanized areas—multi-sensoral approaches using remotely sensed data for risk assessment. Nat Hazards Earth Syst Sci 11: 431–444, Copernicus Publications

The World Bank Group AFTUW (2012) Municipal ICT capacity and its impact on the climate change affected urban poor. The Case of Mozambique

Wang W, Yang X, Yao T (2011) Evaluation of ASTER GDEM and SRTM and their suitability in hydraulic modeling of a glacial lake outburst flood in southeast Tibet, Hydrol Process 26(2):213–225, Whiley Online Library

...

...

Chapter 12
Flood Risk Due to Heavy Rains and Rising Sea Levels in the Municipality of Maputo

Enrico Ponte

Abstract This article assesses flood risk due to heavy rains and sea level rise in the municipality of Maputo (Mozambique) (1.1 million inhabitants in 2007, 346 sq km). The risk assessment methodology took into consideration four factors (hazard, vulnerability, exposure, and adaptation) using the formula $R = (H \times V \times E)/A$, where R = risk, H = hazard, V = vulnerability, E = exposure, A = adaptation. The four factors are measured using different indicators (rainfall, return periods of heavy rains and extreme tides, nature of the soil, density and poverty of population, etc.). The study uses photo interpretation of high-resolution satellite images, population figures, and terrain data, analyzed through the use of open source GIS software and on-site information. Eleven flood prone areas are identified: eight threatened by heavy rains and three by sea level rise. According to these results, an estimated 7 % of the population of Maputo lives in areas (106 ha) where the flood risk is greater than 6 on a scale of 1–10.

Keywords Climate change · Floods risk assessment · Risk modelling · GIS · Maputo

12.1 Introduction

In the last 30 years the Earth has suffered increasingly frequent and intense natural disasters (EM–DAT 2012; World Bank 2011). The number of people affected by these disasters averaged 147 million per year during 1981–1990, which increased to 211 million per year during 1991–2000 and 246 million in 2001–2010.

People, assets, and natural systems increasingly suffer the effects of these natural disasters (Botzen et al. 2010).

E. Ponte (✉)
Interuniversity Department of Regional and Urban Studies and Planning, Politecnico di Torino, Viale Mattioli 39, 10125 Turin, Italy
e-mail: enrico.ponte@polito.it

S. Macchi and M. Tiepolo (eds.), *Climate Change Vulnerability in Southern African Cities*, Springer Climate, DOI: 10.1007/978-3-319-00672-7_12, © Springer International Publishing Switzerland 2014

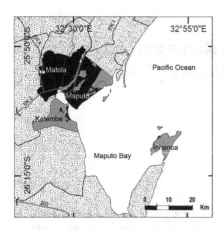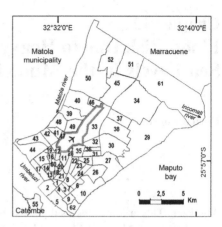

Fig. 12.1 On the *left*, Matola and Maputo municipalities, with suburban districts of Inhanca and Katembe. Built-up area (*black*) and undeveloped area (*grey*); on the *right*, the subdivision of Maputo into *bairros* (map by E. Ponte)

Today, flooding is responsible for two-thirds of the people affected and causes over one-third of the total estimated economic damage (Pilon 2003). Increased flooding due to climate change is affecting increasingly urbanized areas, which threatens economic growth, and requires an integrated approach to risk management (Apel et al. 2009; Shi et al. 2005).

This study assesses flood risk due to heavy rains and sea level rise in Maputo, the capital of Mozambique. Over a million people live in this 346 km² urban area. In particular, this study focuses on six city districts (KaMpfumo, KaChamanculo, KaMaxakeni, KaMubukwana, Ka Mavota, and Katembe—see Chap. 13, Fig. 13.1 for district boundaries) and their 54 *bairros* (neighborhoods), and excludes Matola and the less populated district of Inhaca (Fig. 12.1).

In contemporary society, population increase and ageing, economic development, urbanization, industrialization, and deforestation make urban settlements more vulnerable to natural risk (Takeuchi 2006). The term *natural risk* includes all events which create a

probability of harmful consequences or expected losses (deaths, injuries, property, livelihoods, economic activity disrupted or environmental damaged [*sic*]) resulting from interactions between natural or human–induced hazards and vulnerable conditions (UN-ISDR 2004).

In the last 20 years numerous studies have been carried out on the effects of natural risks in different urban regions (Pelling 1997; Bankoff 2003; Aragon-Durand 2007; Dutta et al. 2005; Thieken et al. 2007; Ali 2007). Others have focused on specific hazards such as floods (Duclos and Isaacson 1987; Legome et al. 1995; Jonkman and Kelman 2005).

In developing countries where many floods are caused by natural risks aggravated by insufficient development in affected communities, the vulnerability of the population living in areas exposed to danger is a challenge for risk management. Vulnerability is defined as

the condition determined by physical, social, economic, and environmental factors or processes, which increase the susceptibility of a community to the impact of hazards (UN-ISDR 2004).

Apart from vulnerability, the literature identifies two additional factors that contribute to natural risk: hazard and exposure (Crichton 2002; ADRC 2005; Alexander 2000).

In the present assessment, a fourth factor was included in the methodology used to calculate risk for Maputo: adaptation. Adaptation to climate change is defined as

the adjustment in natural or human systems in response to actual or expected climatic stimuli or their effects, which moderates harm or exploits beneficial opportunities (IPCC 2001, 982).

Adaptation can be effective in three different ways: it can mitigate the level of vulnerability, reduce exposure and reduce the hazard.

To calculate risk we used the formula (adapted from Davidson 1997):

$$R = \frac{(H \times V \times E)}{A}$$

The four factors (H, V, E, A) were measured using indicators such as rainfall, return periods of heavy rains and extreme tides, and the surface area of hydrographic sub-basins covering the areas exposed to floods. These indicators interact in a very complex way (Smith 1994; Alkema 2003), and how components are evaluated and to which risks they are considered to contribute vary between disciplines (Roberts et al. 2007).

The indicators chosen for Maputo depended on existing data for population, poverty, and rainfall, complemented by information gathered in the field during interviews with district and *bairro* officials (adaptation measures); additional information was obtained through photo-interpretation of a GeoEye satellite image dated September 2010 (bare land, impervious surface rate, municipal dump buffer, tarmac roads).

Combining the multiple components of risk requires a quantitative approach (Roberts et al. 2007) and results of the assessment of flood risk due to heavy rains and extreme tides are presented below. The methodology used to identify risk areas is illustrated, followed by the assessment of hazards, vulnerability, and exposure. The indicators and criteria chosen as well as their weighting are explained. Finally, adaptation is briefly assessed (a more detailed analysis of this factor can be found in Chap. 11).

Although risk assessment focused on all 11 flood prone areas, this article will only report on the area with the highest flood risk values due to heavy rains and the area where there is greatest flood risk due to sea level rise. Results indicate that 7 % of the population in Maputo lives in high-risk areas and that population density and poverty are the indicators with the greatest impact on risk values.

12.2 Identification of Areas with Greatest Flood Risk and Hazards

Flood hazard is defined as the probability of occurrence of a potentially damaging flood event of a certain magnitude in a given area within a specific period of time (Crichton 2002; Kron 2005). The existing literature contains a variety of methodologies for studying hazards.[1]

In the present study, 11 flood prone areas were identified: eight due to heavy rains (A–H) and 3 due to sea level rise (I–K) (Fig. 12.2).

The identification of flood prone areas requires a hydrographic model to simulate the event. In the present case, the municipality's orography was used to identify the areas of Maputo located in the lowest part of the hydrographic sub-basins (see Chap. 11).

12.2.1 Flood Prone Areas Due to Heavy Rains

The eight areas identified as flood prone due to heavy rains varied in size from 68 to 1.383 ha (Table 12.1). Apart from their size, the areas diverged in several ways. Those to the south (E, F, and G) were occupied by formal and informal housing, services, and industries. The areas to the west (B, C, and D) along the river Matola contained informal settlements, cultivated land, and green space, while areas A and H were occupied only by residential housing. There were also differences in the population density of the eight areas.

Hazard (H) is determined by two factors. The first is the quantity of rain in relation to the size of the stormwater drainage system. Rainfall data for the city of Maputo between 1960 and 2006 show that heavy rains in excess of 200 mm have a 15 year return period (see Chap. 7). However, the stormwater drainage system is designed to collect only 100 mm of rain, and will therefore be 100 % over capacity when faced with 200 mm of rain.

The second factor is the surface area of the hydrographic sub-basin of each of the eight flood prone areas due to heavy rains. The bigger the sub-basin, the greater the amount of water that will collect at the bottom of it due to run off.

Therefore, the H value of each area is determined by the ratio between the surface area (in ha) of the flood prone areas (FPA) and the size of its hydrographic sub-basin (Table 12.2) multiplied by 2, which is the ratio between the 200 mm

---

[1] The United Nations Development Program (UNDP) together with the European Commission have initiated a study in the Philippines to define guidelines for including natural disaster risk reduction efforts in development planning processes. In particular, regarding the analysis of hazards, the guidelines envisage dividing the territory according to the intensity or frequency of occurrence and thereby define levels of susceptibility. This is intended to encourage the creation of hazard maps (UNDP 2008).

Fig. 12.2 Maputo. Floodable areas due to heavy rains (*A–F*) according to USGS, and flood prone areas due to sea level rise (*I–K*) (GeoEye photo interpretation, see Chaps. 10 and 11; map by E. Ponte)

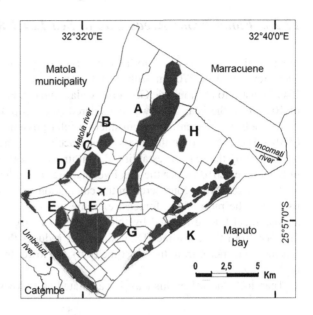

Table 12.1 Flood prone areas of Maputo

Flood prone areas

	A	B	C	D	E	F	G	H	I	J	K	Σ	Maputo
ha	1383	127	192	112	117	778	79	68	1423	256	1197	5732	22900
%	24.1	2.2	3.3	2	2	13.6	1.4	1.2	24.8	4.5	20.9	100	–
%	–	–	–	–	–	–	–	–	–	–	–	25	100

Table 12.2 Surface area and watershed dimensions by flood prone area (FPA)

Dimensions of the watershed

FPA	A	B	C	D	E	F	G	H	Σ
Surface area (ha)	1383	127	192	112	117	778	79	68	2858
Watershed (ha)	2702	596	736	277	304	1364	116	253	6348
W/SA	2	4.7	3.8	2.5	2.6	1.7	1.4	3.7	2.2

heavy rain and the rain for which the stormwater drainage system was designed (100 mm):

$$H = 2 \times \frac{\text{hydrographic sub-basin}}{\text{flood prone area}}$$

The ratio between the hydrographic sub-basins and the flood prone area varies between 1.4 and 4.7.

12.2.2 Flood Prone Areas Due to Sea Level Rise

Identification of areas prone to flooding caused by rising sea levels was premised an extreme tide of 270 cm with a 100 year return period, as established by the INGC (Instituto Nacional de Gestão de Calamidades). Furthermore, a sea level rise of 100 cm in the year 2100 was considered (see Chap. 8).

As such, the present study considers a total possible sea level rise of 370 cm (270 + 100 cm), and on that basis the surface area of the municipality that would be flooded by the sea was established. Within that area, the built areas that existed in September 2010 as per the GeoEye satellite image have been identified.

Two factors were considered in the calculation of flood hazard due to sea level rise: first, the ratio of the forecast extreme tide of 270 cm to the highest annual astronomical tide (estimated by INGC to be 178 cm), which is 1.52; second, the distance from the coastline (DC), since this reduces the impact of flooding. The greater the distance from the coastline (km), the lower the amount of water in case of flooding.

Therefore, hazard in this case is calculated as follows:

$$H = \frac{1.52}{DC}$$

12.3 Vulnerability

The term vulnerability has long been used in literature about natural disasters (Gilbert 1995; Hewitt 1983, 1997), but it became more important when people began to talk about growth and global change (Dow 1992; Dow and Dowing 1995). The IPCC Fourth Assessment Report (IPCC 2007) emphasizes that developing countries are more vulnerable to climate change due to a lack of institutional capacity.

12.3.1 Methodology

Vulnerability has been described by several authors in general terms, but rarely has it been measured in a specific urban context in Sub-Saharan Africa.

As regards social sciences, attempts have recently been made to quantify certain aspects of the problem using anthropological and economic parameters in order to compare different places and time periods (Alwang et al. 2001).

In the literature on vulnerability, it is divided into physical and social vulnerability (Adger 2006; Douglas 2007). Vulnerability is multi-faceted, and its main elements can be described as physical (the natural and built environment), systemic, social/community/institutional, and economic. Each element influences the others (Menoni et al. 2012).

Vulnerability refers to a specific place and time. As a result, specific indicators can be used only if the scale is small (Cutter et al. 2003).

12.3.2 Identification and Weighting of Indicators

To assess the vulnerability of Maputo, this study uses indicators valid not only for heavy rains but also extreme tides.

The following indicators are used to measure vulnerability in flood prone areas: nature of the soil, impervious surfaces (built up areas), lack of tarmac roads, lack of tree cover, poverty of resident population, and proximity to dumps. An example of how indicators were recorded for flood prone area F (see Fig. 12.2) is provided in Table 12.3.

At least two more indicators of vulnerability should have been added: unsafe sanitation and hygiene services (latrines) and access to drinking water from open-air wells which may be polluted by contaminated flood water. This study does not consider these factors because the information available regarding latrines and wells was outdated (2003) and limited to two districts (SEED 2010).

Having established the indicators of vulnerability, the first step is to measure them within each flood prone area.

The second step was to establish the weights of every indicator. If the sum of all the indicators gives a value of less than 1, then vulnerability is mitigated. If it remains 1, then vulnerability is the same. If it is greater than 1, vulnerability increases.

The six indicators used and the weighting methodology are described below:

1. Nature of the soil. Rainwater infiltration was zero in clay terrains along the Costa do Sol and the left bank of the Umbeluzi and Infulene rivers, which occupy 11 % of the municipality (including the districts of Katembe and Inhaca). All the other soils are very sandy and this facilitates infiltration. The sand layer resting on a clay bed ranges from 45 to 75 m (Geological Map of Maputo 1:50,000). Nevertheless, some sandy surfaces (e.g. the Congolote formation) are sometimes covered by a layer of clay (e.g. the Machava formation) that reduces permeability, making water infiltration impossible (Isidro and Vicente 2004). The following values were attributed to the two different kinds of soils found in the municipality: 0.1 for sand and 1 for clay (because it is, however, slightly impervious).

2. Impervious surfaces. Water collected on impervious surfaces (roofs, floors) runs into permeable interstices. When there are more impervious than permeable surfaces, water tends to stagnate, sometimes covering large areas. Since Maputo has a patchy stormwater drainage system, the impervious surface rate measures this factor of vulnerability. This information was taken from a high

Table 12.3 Layers of vulnerability indicators for area F

Area (Fig. 12.2)	No of layers	Serial letter	Soil	Imperv. surf.	Roads	Tree cover	Poverty	Dump	$\sum$ indices
F	2	a	0	1	0	0	1	0	2
F	2	b	0	1	0	1	0	0	2
F	3	a	0	0	1	1	1	0	3
F	4	a	0	1	1	1	1	0	4
F	2	a	0	1	0	0	1	0	2
F	3	a	0	0	1	1	1	0	3

resolution GeoEye satellite image (2010). Each exposed area was divided into territorial units with homogeneous settlements. One hectare in each unit was used as a sample and measurements were taken of its impervious surface rate (Fig. 12.3). The impervious surface rates ranged from 9 to 67 %.

The territory in question was divided into areas of homogenous settlement, and the impervious surface rate for each area was calculated. A value of 1.1 was attributed to areas with an impervious surface rate of less than 50 %, and a value of 2 to areas with a rate of 50 % or more.

3. Roads. The surface of a road affects its the ability to absorb water. In Maputo, most roads are unsurfaced, based on the GeoEye satellite image dated September 2010 and additional data collected during on-site inspections (Fig. 12.4). During these inspections we found that asphalt or concrete block roads do not always have a suitable drainage system. In such cases, water runs off the road surface. With some exceptions, unsurfaced roads have no drainage so they are unable to quickly drain off water. As such, asphalt roads were assigned a value of 0 while unsurfaced roads were given a value of 1.

4. Tree cover. Tree cover delays and reduces the amount of rainwater that reaches the ground and therefore reduces vulnerability. Based on the high-resolution GeoEye image and on-site inspections, the municipal areas without tree cover were identified. In particular, two classes and their relative values were defined: bare ground was assigned a value of 1, and soil with trees was assigned a value of 0.5.

5. Poverty. Vulnerability is directly related to poverty. Métier's 2007 estimation of poverty in the *bairros,* conducted for the World Bank, ranged from a low of 20 % in the *bairro* of Malhangalene A to a high of 74 % in the *bairro* of Albazine. The parameters used in that study were housing type, access to drinking water, energy source, level of education, and type of employment. It's important to note that the *bairros* with a lower poverty index (less than 30 %) were found in urban district 1 (KaMpfumo), while the 8 areas identified in the present study as prone to flooding due to heavy rains have a poverty index ranging from 30 to 60 %. Métier's poverty percentage was used for each *bairro* (e.g. bairro Chamanculo C = 56 % = 0.56).

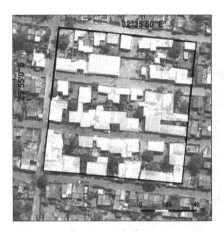 

Fig. 12.3 Roof cover density: 45 % (*left*) and 43 % (*right*) (E. Ponte)

Fig. 12.4 *Left* Maputo, Mafalala bairro, Nov. 2011. People walking along an unpaved flooded road; *Right* Maputo, Costa do Sol bairro, Rua Engenheiro Santos Resenoe, Nov. 2011. Paved road with a drainage canal (E. Ponte)

6. Dump sites. The municipal dumpsite is located in bairro Hulene B. The dumpsite was created 45 years ago on an abandoned fresh water lagoon. The dump now covers 25 ha and is used for all kinds of rubbish, including domestic, sanitary, industrial, and commercial waste (Vicente et al. 2006). The garbage mound, varying from 5 to 15 m in height, is located less than 10 m from adjacent houses, and there is no protection/retaining wall (IBAM 2008). Heavy rains will make the surrounding area even more vulnerable due to the run off from the mountain of trash and the percolation of contaminated water in the water table supplying the wells. No standard buffer zones are specified in the regulations regarding dumps. Based on Italian regulations, we considered that the effects of the rubbish would be harmful to the health of anyone residing within 500 m of the dump. Given the extreme hazard created by this situation, we assigned a value of 10 to the 500 m wide buffer zone.

12.4 Calculating Vulnerability

Having established coefficients for the six indicators, the vulnerability index was calculated for each plot of land in the eight areas prone to flooding due to heavy rains.

Four of the six indicators were considered elements of physical vulnerability because they involved the terrain: tree cover, impervious area, type of soil, and road surface. The severity of physical vulnerability is then multiplied by the social vulnerability factor (poverty). In the case of heavy rains, an extremely poor population may be unable to move to new accommodations and may not have the capacity to cure illnesses caused by flooding (malaria, diarrhea, leptospirosis, etc.) This is why the poverty indicator is considered a magnifier. Finally, proximity to the dump is added as a separate, additional element.

In conclusion, this is the formula; below we have provided the variability range of each indicator:

$$V = (\text{Soil} + \text{Impervious Surface} + \text{Roads} + \text{Tree Cover}) \times \text{Poverty} + \text{Dump}$$
$$V = ((0.1 \div 1) + (1.1 \div 2) + (0 \div 1) + (0.5 \div 1)) \times (0.2 \div 0.75) + (0 \div 10)$$

12.5 Exposure

Exposure is defined as

> people, property, systems, or other elements present in hazard zones that are thereby subject to potential losses (UN-ISDR 2009).

The indicators to consider when determining exposure are population density and the economic and strategic importance of the exposed elements (infrastructure, services, industries).

The population density of Maputo was determined on the basis of the 2007 population census. Using the *bairro* as the basic data unit, the entire municipality can be divided into 62 units. To more accurately calculate the density of a specific unit, the area of any unbuilt sites of more than 1 ha was subtracted from the surface area of the unit. Unbuilt areas were identified through photo interpretation of GeoEye satellite images (2010) (Fig. 12.5).

According to those calculations, Maputo has a mean density of 46 inhabitants per hectare (i/ha): density peaks occur in Chamanculo (294 i/ha) and Maxaquene (291 i/ha); meanwhile, the lowest values occurred in the suburban areas of Bagamoio (31 i/ha) and Katembe (2 i/ha). However, some suburban areas are densely inhabited, for example Hulene A (220 i/ha) (subdivision 32) and 25 de Junio B (148 i/ha) (subdivision 49) (see Fig. 12.1).

Fig. 12.5 Built-up area (*grey*) used to calculate population density. *White areas* are undeveloped plots larger than 1 ha (GeoEye 2010; map by E. Ponte)

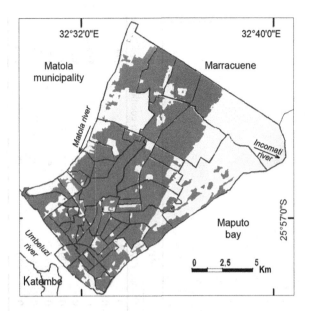

According to study results, only a few densely populated areas are flood prone (Chamanculo, Minkadjuine, Mafalala, and Maxaquene). Nevertheless, the less densely populated northern *bairros*, which are also flood prone, are experiencing rapid demographic growth. Moreover, the situation illustrated in Fig. 12.5 refers to 2010, and may already have worsened.

On-site inspections and interviews revealed that official markets and industrial zones in flood prone areas were rarely affected because they were built on foundations. Meanwhile, most informal markets and some schools were flooded on a regular basis. A complete exposure analysis should include this data.

12.6 Adaptation

The value assigned to adaptation, defined in detail by Tiepolo (see Chap. 13), is based on the number of flood impact reduction measures implemented. The literature divides such measures into structural and non-structural (FIFMTF 1992). The former are visible physical structures (stormwater drainage canals, raised foundations to protect buildings, sea walls), which are generally quite expensive. Non-structural measures (early warning systems, sanitary education, event simulations) tend to cost less than structural measures and are not visible.

Adaptation measures were analyzed in detail for the four districts identified as regularly exposed to flooding: Katembe, KaChamanculo, KaMaxakeni, and KaMavota (see Chap. 13). The data collected on adaptation measures during interviews and on-site inspections in those areas were georeferenced on a map.

Table 12.4 Maputo. Sample risk calculation index for some areas outlined in Fig. 12.6

Bairro	ha	Hazard		Vulnerability						Exp.	Adap.	Risk
		Rain	Watershed	Poverty	Soil	Imperv. surf.	Roads	Tree cover	Dump	Population density	Adaptation coefficient	Risk value
Aeroporto A	30.98	2	1.75	0.49	0.5	2	1	0.5	0	143.5	0.25	4.06
Aeroporto A	10.45	2	1.75	0.49	0.5	2	0	1	0	143.5	0.25	3.04
Aeroporto A	2.71	2	1.75	0.49	0.5	2	1	1	0	143.5	0.25	5.08
Aeroporto B	8.93	2	1.75	0.56	0.5	2	1	0.5	0	179.5	0.25	5.25
Aeroporto B	2.35	2	1.75	0.56	0.5	1	1	1	0	179.5	0.25	6.53
Mafalala	24.86	2	1.75	0.53	0.5	1	1	1	0	192.7	0.5	2.77
Minkadjuine	3.2	2	1.75	0.51	0.5	2	1	0.5	0	189.4	0.25	6.75
Munhuana	30.54	2	1.75	0.53	0.5	1.1	1	1	0	70.82	0.25	2.04

Fig. 12.6 Maputo. Risk areas within the flood prone area F with a risk score ranging from 6.6 to 0.1. Risk score: very high (*black*), high (*grey*) and moderate (*white*) (E. Ponte)

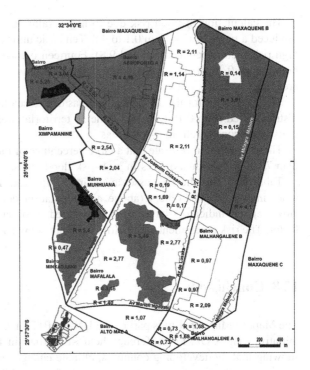

Not all adaptation measures have the same impact on risk reduction, so each measure was weighted. Once all the measures in the districts in question had been identified, their weighting produced the following results: Katembe 0.25, KaChamanculo 0.25, KaMaxakeni 0.5, KaMavota 0.75.

12.7 Risk

Risk was analyzed in terms of the four risk factors—hazard, vulnerability, exposure and adaptation—using the following formula:

$$R = (H \times V \times E)/A$$

Heavy Rains:

R={[2 * (sub - bas/area)] × [(roads + soil + tree cover + imperv. surf.) × pov + dump] × (pop dens.)}/(adapt. meas.)

R={[2 * (1.9 ÷ 4.7)] × [(1 + 0.1 ÷ 1 + 0.5 ÷ 1 + 1.1 ÷ 2) × 0.05 ÷ 0.65 + 0 ÷ 10] × (294 ÷ 2)}/(0.25 ÷ 0.75)

Sea level rise:

R={[1.52 * (dist.c)] × [(roads + soil + tree cover + imperv. surf.) × pov + dump] × (pop dens.)}/(adapt. meas.)

R={[1.52 * (0 ÷ 2)] × [(1 + 0.1 ÷ 1 + 0.5 ÷ 1 + 1.1 ÷ 2) × 0.05 ÷ 0.65 + 0 ÷ 10] × (294 ÷ 2)}/(0.25 ÷ 0.75)

The calculation of flood risk due to heavy rains for 100 basic territorial units produced a range of risk from 0.1 to 9.8. Ten basic units (37 ha) have an extremely high risk factor (between 6.0 and 9.8). Fourteen other basic units are considered high-risk (3.0–5.9). The 10 extremely high-risk units represent 10 % of floodable areas due to heavy rains.

The risk factor for the 10 floodable areas due to sea level rise in the KaMavota district varied from 0.19 to 9.38. Of the 36 territorial areas analyzed, 13 territorial units were considered high-risk areas.

These results demonstrate that a large percentage of the population (slightly less than 70,000 individuals) lives in high-risk flood areas.

Table 12.4 and the risk assessment maps illustrate how some indicators are more important than others (above all, population density and poverty index). However, all indicators have to be considered in order to obtain a reliable risk factor. The risk assessment results for area F are shown as an example in Fig. 12.6.

12.8 Conclusions

The Maputo Municipality is part of the UN-HABITAT *Cities and Climate Change* initiative. The COP 17 conference held in Durban in late 2011 presented data showing that, in developing countries, climate change and related natural hazards impact coastal cities the most. Sixty percent of the world's population lives in low-lying coastal areas (less than 10 m.a.s.l.). Although these areas represent only 2 % of surfaces all over the world, they support 10 % of the population, 80 % of which live in cities. The UN ISDR project, Making Cities Resilient: *My city is getting ready*, is planning important studies involving a large number of coastal cities, proof of the relevance and importance of this study.

The goal of the present study is to prepare a tool that the Maputo Municipality could use to assess flood risk. As in many large cities, very little data exists about the population, climate, settlements, and infrastructure south of the Sahara. Although the limited availability of data affected the methodology used in the study, a useful tool has been developed with the following characteristics:

- Simple. The risk index is visualized on coded maps showing a number corresponding to the numeric value obtained. The different grey hatched areas on the map show the risk factor (black = very high risk, white = low risk);
- Accurate. The areas at risk are shown on maps and are immediately identifiable thanks to appropriate scales and names of the road network;
- Transferable. The data was processed using open source GIS, which allows data to be exchanged free of charge with other users;
- Easily updated. Risk area maps can be updated by changing the input of the layers.

Future research into this area should take into account the fact that floods due to heavy rains or sea level rise can coexist in Maputo. So far, these two sources of risk have been considered only separately. As such, a multi-risk assessment should be developed, one that draws on the input of several different disciplines (EMA 2002).

This study assessed both types of risk on the basis of indicators—namely population density, poverty, construction and adaptation measures—that are constantly evolving in Maputo. Future application of this methodology will therefore require a shift from use of the risk assessment tool developed by this study to ongoing use of a risk monitoring tool.

References

Adger WN (2006) Vulnerability. Glob. Environ. Change 16:268–281

ADRC (2005) Total disaster risk management—good practices, report. Asian Disaster Reduction Center, Kobe

Alexander D (2000) Confronting catastrophe: new perspectives on natural disasters. Oxford University Press, Oxford

Ali AMS (2007) September 2004 flood event in southwestern Bangladesh: a study of its nature, causes, and human perception and adjustments to a new hazard. Nat Hazards 40:89–111

Alkema D (2003) Flood risk assessment for EIA; an example of a motorway near Trento, Italy. Studi Trentini di Science Naturali Acta Geologica 78:147–153

Alwang JPB, Siegel J et al (2001) Vulnerability: a view from different disciplines. Social protection discussion paper series no. 0115. Social Protection Unit, Human Development Network, The World Bank, Washington

Apel H, Aronica GT et al (2009) Flood risk analyses—how detailed do we need to be? Nat Hazards 49:78–98

Aragon-Durand F (2007) Urbanization and flood vulnerability in the peri-urban interface of Mexico city. Disasters 31(4):477–494

Bankoff G (2003) Constructing vulnerability: the historical, natural and social generation of flooding in metropolitan Manila. Disasters 27:224–238

Botzen W, Van den Bergh J, Bouwer L (2010) Climate change and increased risk for the insurance sector: a global perspective and an assessment for Netherlands. Nat Hazards 52:557–598

Crichton D (2002) UK and global insurance responses to flood hazard. Water Int 27(1):119–131

Cutter SL, Boruff BJ, Shirley WL (2003) Social vulnerability to environmental hazards. Soc Sci 84(2):242–261

Davidson R (1997) An urban earthquake disaster risk index. The John A. Blume Earthquake Engineering Center, Department of Civil Engineering, Report no. 121. Stanford University, Stanford

Dow K (1992) Exploring differences in our common future(s): the meaning of vulnerability to global environmental change. Geoforum 23:417–436

Dow K, Dowing TE (1995) Vulnerability research: where things stand. Hum Dimensions Q 1:3–5

Douglas J (2007) Physical vulnerability modeling in natural hazard risk assessment. Nat Hazards Earth Syst Sci 7:283–288

Duclos P, Isaacson J (1987) Preventable deaths related to floods, letters to the editor. Am J Public Health 77(11):14–74

Dutta D, Khatun F, Herath S (2005) Analysis of flood vulnerability of urban buildings and populations in Hanoi, Vietnam. Seisankenkyu 57(4):338–342

EMA (2002) Disaster loss assessment guidelines. Part III emergency management practice, vol 3, Guide 11

EM–DAT (2012) The international disasters database. WHO collaborating centre for research on the epidemiology of disasters. Université Catholique de Louvain, Louvain

FIFMTF (1992) Floodplain management in the United States: an assessment report, vol 2: full report. L. R. Johnston Associates, Marietta

GeoEye, satellite image, date of purchase 8 September 2011, Product Work Order Number: SG00047170_001_001780279; date of acquisition 26 September 2010, ID 201009260752582160303160l826

Gilbert C (1995) Studying disaster: a review of the main conceptual tools. Int J Mass Emerg Disasters 13(3):231–240

Hewitt K (1983) Interpretations of calamity: from the viewpoint of human ecology. Allen and Unwin, London

Hewitt K (1997) Regions of risk–a geographical introduction to disasters. Addison Wesley Longman Limited, Harlow

IBAM (2008) Urbanização e desenvolvimento municipal em Moçambique. Capitulo: Gestão de resíduos sólidos. Relatório técnico final, junho, UN-Habitat, The World Bank, Washington

IPCC (2001) Climate change 2001: impacts, adaptation and vulnerability. Contribution of working group II to the fourth assessment report of the intergovernmental panel on climate change. Cambridge University Press, Cambridge

IPCC (2007) Climate change 2007: impacts, adaptation and vulnerability. Contribution of working group II to the fourth assessment report of the intergovernmental panel on climate change. Cambridge University Press, Cambridge

Isidro RM, Vicente EM (2004) Mozambique: Maputo, a geo-environmental hazard prone city, in disaster reduction in Africa. ISDR (International Strategy for Disaster Reduction) Informs 3:20–22

Jonkman SN, Kelman I (2005) An analysis of the causes and circumstances of flood disaster deaths. Disasters 29(1):75–97

Kron F (2005) Flood risk = Hazard * Values * Vulnerability. Water Int 30(1):58–68

Legome E, Robins A, Rund DA (1995) Injuries associated with floods: the need for an international reporting scheme. Disasters 19(1):50–54

Menoni S, Molinari D, Parker D, Ballio F, Tapsell S (2012) Assessing multifaceted vulnerability and resilience in order to design risk-mitigation strategies. Nat Hazards. doi:10.1007/s11069-012-0134-4

Métier P (2007) Cidade de Maputo. Avaliação e mapeamento da pobreza, In-site 9, Banco Mundial, Washington

Pelling P (1997) What determines vulnerability to floods: a case study in Georgetown, Guyana. Environ Urban 9:203–226

Pilon PJ (2003) Guidelines for reducing flood losses, report. UN DESA (Department of Economic and Social Affairs), WMO (World Meteorological Organization), Geneva

Roberts N, Nadim F, Kalsnes B (2007) Quantification of vulnerability to natural hazards. Georisk 3(3):164–173

SEED (2010) Formulation of an outline strategy for Maputo city C wide sanitation planning. Final report, 15 September, Maputo

Smith D (1994) Flood damage estimation: a review of urban stage-damage curves and loss function. Water S Afr 20(3):231–238

Shi P, Ge Y, Yuan Y, Guo W (2005) Integrated risk management of flood disaster in metropolitan areas of China. Water Resour Dev 21(4):613–627

Takeuchi K (2006) ICHARM calls for an alliance for localism to manage the risk of water-related disasters. In: Tchiguirinskaia I, Thein KNN, Hubert P (eds) Frontiers in flood research, IAHS (International Association of Hydrological Science), Red Book Series 305, London

Thieken AH, Kreibich H, Muller M, Merz B (2007) Coping with floods; preparedness, response and recovery of flood affected residents. J Hydrol Sci 52(5):1016–1037

UN-ISDR (2004) Terminology: basic terms of disaster risk reduction. International strategy for disaster reduction secretariat, Geneva. http://www.unisdr.org/eng/library/lib-terminology-eng%20home.htm

UN-ISDR (2009) Terminology on disaster risk reduction. International strategy for disaster reduction secretariat, Geneva. http://www.unisdr.org/eng/terminology/terminology-2009-eng.html

UNDP (2008) Mainstreaming disaster risk reduction in subnational development land use/physical planning in the Philippines, Geneva

Vicente EM, Jermy CA, Schreiner HD (2006) Urban geology of Maputo, Mozambique, IAEG paper no 338. The Geological society of London, London

World Bank (2011) The little data book on climate change. The World Bank, Washington

Chapter 13
Ongoing and Future Flood Adaptation Measures in the Municipality of Maputo

Maurizio Tiepolo

Abstract This chapter presents the adaptation baseline for flooding caused by extreme rains, sea level rise, and high tides brought on by climate change in the municipality of Maputo (1.1 million inhabitants in 2007, 347 km^2). Adaptation measures were ascertained over the 57.4 km^2 exposed to regular flooding through interviews with district and neighborhood officers and on-site inspections. The size of the areas exposed to flooding (16 % of the administrative surface), their dispersion in 21 fragments and the fact that many measures are not detectable by means of surveys has complicated the work. The importance of adaptation is defined for each exposed area on the basis of the physical, economic, social, and health consequences of its absence and is expressed as a figure that can be used in risk equations. A focus group set up with members of the environmental directorate of the municipality of Maputo identified short, medium, and long-term measures, prioritized those measures according to five criteria, and discussed the mainstreaming of adaptation in planning tools.

Keywords Climate change · Flood risk · Assessment of adaptation measures · Adaptation planning · Maputo

13.1 Introduction

As mentioned in previous chapters, two physical events that are influenced by climate change were investigated in Maputo: extreme rains, i.e. daily rainfall greater than 200 mm (see Chap. 9); and the combination of extreme tides and sea level rise (see Chap. 10). This chapter identifies the adaptation baseline (Levina

M. Tiepolo (✉)
Interuniversity Department of Regional and Urban Studies and Planning, Politecnico di Torino, Viale Mattioli 39, 10125 Turin, Italy
e-mail: maurizio.tiepolo@polito.it

S. Macchi and M. Tiepolo (eds.), *Climate Change Vulnerability in Southern African Cities*, Springer Climate, DOI: 10.1007/978-3-319-00672-7_13, © Springer International Publishing Switzerland 2014

and Tirpak 2006) as well as the future measures that should be prioritized and mainstreamed in local planning so as to reduce flooding risk.

The need for adaptation to urban climate change in Mozambique is significant and cannot be addressed by international aid alone, not even in regions that receive significant funding from international donors, such as the Sahel (Tiepolo 2012). To achieve adaptation, aid should instead be used to support local governments in finding and using their own resources more efficiently (Braccio and Tiepolo 2013), hence the need to identify areas of greatest risk, prioritize actions, distribute those actions over time, and ensure they are mainstreamed in planning tools.

Since no previous assessment of adaptation to floods and related physical planning adjustments had been conducted for Maputo (United Nations Habitat 2010), this chapter also investigates future adaptation measures for the city. The literature dealing with large African cities prone to heavy rains brought on by climate change identifies drainage as a typical adaptation measure (Lwasa 2010: 167; Folorunsho and Omojola 2009; Karley 2009; République centrafricaine 2009; IMWI 2013). Although drainage work is fundamental, the importance of alternative devices that can be managed jointly to encourage the infiltration and storage of water (filtering trenches and wells, infiltration basins) has been recognized as well (Dasylva 2009). There is still general agreement on the inevitability of resettlement procedures affecting residents in areas prone to regular flooding.

13.2 Adaptation Measures Defined

The equation used to calculate risk includes not only hazard, vulnerability, and exposure, but also adaptation:

$$R = (H * V * E)/A \text{ (adapted from Davidson 1997)}$$

The term *adaptation to climate change* indicates

the adjustment in natural or human systems in response to actual or expected climatic stimuli or their effects, which moderates harm or exploits beneficial opportunities. (IPCC 2001)

There are two kinds of adaptation measures: structural and non-structural (FIFMTF 1992). As the former involve public works, they are visible: rainwater drainage canals, platforms or stilts used to raise homes above high water levels, new settlements built to house the inhabitants of areas prone to flooding, and sea walls, stone walls, embankments, and raised door thresholds to protect against high tides. These are expensive improvements whose specific geographic location can be determined through on-site inspections, a long but feasible operation.

Non-structural measures are more numerous. These include early warning devices alerting residents to the increased danger of flooding, the education of local people in terms of what should be done should flooding occur, dissemination of information on hygiene, conducting flooding drills in schools and hospitals, formulation of floodable area maps and evacuation maps, and the installation of markers indicating the water level in exposed areas. In other cases, non-structural measures include less obvious activities that often prove to be vital in the event of flooding: installation of electric generators, drinking water tanks for schools and hospitals, the protection of industrial water treatment plants or of warehouses storing dangerous substances, latrines raised above the high water mark, and the protection of drinking water wells against contamination by floodwater. Other measures come into play only after the event, such as vaccination campaigns and the replanting of mangrove forest areas, the latter of which is not always visible due to the slow rate of growth. As such, many of these non-structural adaptation measures are not obvious when carrying out on-site inspections.

13.3 Data Collection Methodology

Adaptation measures in Maputo are urgently needed in the areas most exposed to flooding due to heavy rains (28.6 km^2) and the areas most exposed to flooding due to extreme tides and sea level rise (28.8 km^2), as identified by Braccio (see Chap. 11) and Brandini and Perna (see Chap. 10). Adaptation measures were therefore ascertained for 21 areas of Maputo, covering a total of 57.4 km^2: a large area if one intends to carry out comprehensive inspections.

When ascertaining adaptation, the methods to be used should depend on the conditions of the particular case (Barros et al. 2012).

As regards Maputo, the method used to identify an adaptation baseline (Table 13.1) involved interviews with district and *bairro* (neighborhood) officers working in exposed areas.

Some measures, such as the rainwater drainage network, might have been identified without the need for on-site inspections were it not for the fact that there are no complete, up-to-date maps showing the current level of maintenance.

Therefore, two interviews were conducted in each of the four districts where most of the areas exposed to flooding are located: an in-depth interview with a district officer, followed by a general inspection of the district, and an in-depth interview with the secretary of the neighborhood most at risk, followed by visits to specific sites. Each interview was structured around four topics: the hazards encountered over the previous year; what happened before, during, and after the event; the existing adaptation measures; and the measures that needed to be introduced. When discussing hazards, interviewers specifically mentioned the heavy rains of late January 2011 (200 mm) and 1 November 2011, which, though considered modest, produced immediately visible effects in the most exposed neighborhoods, including the accumulation of standing water.

The only portion of the interview that involved prompts concerned existing measures. Using a photo album depicting 32 typical measures listed in the literature, interviewees were asked to identify the measures that existed in their district/neighborhood and to describe the extent thereof. Interviews took place

Table 13.1 Maputo: adaptation study methodology

Ongoing measures	Information sources	Future measures	Information sources
• Measure identification	• District/neighborhood officer interviews	• Priority criteria	• Maputo directorate for the environment
• Measure weight	• Field visits	• Measure identification	Focus group
• District ranking	• Desk work	• Measure ranking	
▼		▼	
$R = (H*V*E)/A$		6 priority measures	

between 1 and 4 November 2011, and were conducted by the author, assisted by E. Ponte and accompanied by an officer of the environmental directorate of the municipality of Maputo, in districts no. 2 KaChamanculo (neighborhood no. 15 Chamanculo C), no. 3 KaMaxakeni (neighborhood no. 27 Polana Caniço B), no. 5 KaMavota (neighborhood no. 51 Maghoanine B) and no. 6 Katembe (neighborhood no. 55 Guachene) (Fig. 13.1).

Non-structural measures are extremely important, such as early warning systems and the construction of local climatic scenarios for forecasting the return period of heavy daily rains, which are particularly useful in planning drainage work (Magadza 2000). Non-structural measures also include the coping strategies of the poor, such as savings groups, income diversification and the accumulation of assets (Allen et al. 2010). Nevertheless, as regards flooding due to sea level rise, spontaneous, improvised adaptations are believed to have little effect compared to investment done from public sector (Nicholls et al. 2010: 5).

In order to ascertain whether such measures are important in Maputo as well, what the priorities are, and what their short, medium, and long-term distribution could be, we set up a focus group with six officers from the municipality's environmental directorate. A few reflections on this are given at the end of the chapter.

13.4 Existing Adaptation Measures

The adaptation measures currently in place in areas exposed to flooding were identified by showing each interviewee a photo album of 32 typical measures listed in the literature. This allowed researchers to distinguish 18 adaptation measures currently in place in Maputo. The most common measure consisted in the identification of inhabitants affected by flooding in each district, carried out by the *chefe de quarteirão* (the heads of a small number of blocs) following episodes of heavy rainfall. These reports are submitted to District officials, and used to compile a list of flooded *quarteirão* (each *quarteirão* is identified with a progressive numbering system). In the future, it would therefore be possible to build a georeferenced map of the areas flooded after each heavy rainfall.

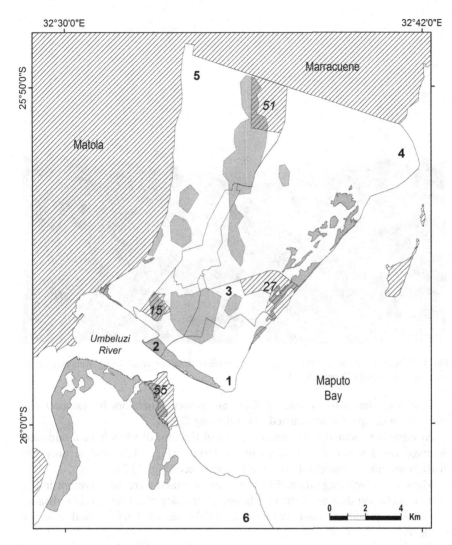

Fig. 13.1 Maputo 2011. Interviewed districts (no. *2, 3, 5, 6*) and neighborhoods (no. *15, 27, 51, 55*); areas exposed to flooding from extreme rain, extreme tides, and sea level rise are highlighted in *grey*

Three of the four districts notify residents when dangerous weather is forecast. In two districts, an early warning is launched via television and radio. However, the information provided is very general and not specific to the situation in each neighborhood or district. In the district of Katembe, for example, a local radio station would be much more effective because it would be closer to the situation and could provide district-specific warnings.

Fig. 13.2 Maputo, 5 Nov. 2011. A primary rainwater drainage canal along Kahunda Street, properly maintained (Tiepolo 2011)

In two of the districts studied, there are power generators in hospitals and vaccination campaigns are carried out following floods.

As regards structural measures, only two of the four districts have a rainwater drainage canal network and maintenance service (Fig. 13.2), and in only two districts are homes equipped with a raised entrance (Fig. 13.3).

Many of the existing adaptation measures identified were only present in one district, including disaster drills in schools, emergency plans for schools, disaster preparation, drinking water tanks in hospitals, raised latrines, and resettled residents.

The following measures were completely absent: formulation of maps of flooded areas and evacuation maps, use of high water markers, accumulation of emergency food reserves prior to floods, implementation of economic incentives discouraging people from constructing buildings in flood-prone areas, dissemination of emergency kits, protection of industrial waste water treatment plants, construction of flood barriers and formulation of emergency plans for industrial areas, instructions on handling dangerous substances, house retrofitting, construction of cement barriers to protect drinking water well heads, dune and mangrove restoration, stabilization of banks and hillside slopes, and the construction of new rainwater drainage systems.

Fig. 13.3 Maputo, KaMaxakeni district, 2 Nov. 2011. The raised entrances of building plots constructed to exclude floodwater (Tiepolo 2011)

13.5 Distribution of Adaptation Measures

The distribution of adaptation measures across the four districts studied varies considerably. In Katembe, on the right bank of the Umbeluzi River, and in KaChamanculo only 4 measures are in place, while 8 measures were identified in KaMaxakeni and 13 were identified in KaMavota.

As regards the flood risk due to tides and rising sea levels, the analysis was limited to the area of KaMavota, where stone barriers and sea walls were found to be in place (Fig. 13.4 and Table 13.2).

The adaptation baseline data was then used to compile a risk map (see Chap. 12) in which all four risk calculation variables (hazard, vulnerability, exposure, and adaptation) were georeferenced. On that basis, a specific risk figure was calculated for each point of the exposed areas, indicating the risk intensity on a scale from 1 to 10. Information regarding adaptation, collected through interviews and site visits, was also included in the map. When the secretary general of a neighborhood identified commonly used adaptation measures without specifying specific zones where they were applied, the map simply indicates such measures are present throughout the entire neighborhood.

Fig. 13.4 Maputo, district 4, 3 Nov. 2011. Sea wall in the Costa do Sol neighborhood (Ponte 2011)

13.6 The Importance of Existing Adaptation Measures

Not all adaptation measures are of equal importance. Each measure was weighted according to the physical, economic, social and health consequences that would occur in its absence (Table 13.2).

Based on site visits to flood-prone areas, it appears that while most structural and non-structural measures can't prevent flooding, the dearth or total lack of such measures correspond to increased severity of the effects of flooding. Since adaptation is the denominator in the risk equation, even at its maximum value of 1 it cannot reduce risk. Rather in a best-case adaptation scenario, a denominator of 1 will have no impact on the overall risk value, while a denominator of less than 1 (e.g. an adaptation score of 0.5) will increase the total risk score.

At this point in the study, the physical, economic, social, and health consequences in the absence of each adaptation measure in case of flooding were considered (Table 13.2).

Physical consequences include the interruption of drinking water and electricity supplies, sewage leakage from latrines, and homes rendered uninhabitable. Health consequences include the spread of diseases such as malaria, diarrhea, leptospirosis, and cholera as well as the interruption of healthcare and delays in emergency services.

Table 13.2 Maputo 2011. The weight of adaptation measures and their presence in the municipal district of KaChamanculo (2), KaMaxakeni (3), KaMavota (5) and Katembe (6)

Adaptation measure	Consequence in measure's absence				Score	District score			
	Physical	Health	Economic	Social		2	3	5	6
Nonstructural									
Early warning	–	Injured	Loss of goods	Poverty	0.075	•		•	
Residents information	–	Diseases		–	0.05	•		•	
School disaster simulation	–	Injured	–	–	0.025		•	•	•
Preparedness training	–	Injured	Loss of goods	Poverty	0.075		•	•	•
Hospital power generator	Outage	No treatment	–	–	0.05			•	•
Exposed residents id.	–	Aid delivery	–	Poverty	0.05	•	•	•	•
School emergency plan	–	Injured	–	–	0.025		•	•	•
Water tank in hospitals	Outage	Hygiene	–	–	0.05			•	
Elevated latrines	Sludge leaks	Diarrhea, cholera	Loss	Poverty	0.1		•	•	•
Vaccination campaign	–	Cholera	of	Poverty	0.075		•	•	•
Hygiene education	–	Diarrhea	income	Poverty	0.075			•	
Mangroves restoration	Erosion Flood	–	Energy loss	–	0.05			•	•
Structural									
Storm water drainage – maintenance	Flood	Malaria	Loss of crop circulation	Poverty medical supply	0.1	•	•		
– extension	Flood	Malaria	Stop	–	0.1		•	•	
Resettlement	Flood	Diseases	Loss	Poverty	0.1			•	•
Stone barrier	Erosion	–	of	Poverty	0.075			•	•
Sea wall	Erosion	Diseases	goods	Poverty	0.1			•	
Total extreme rain					1.075	0.275	0.425	0.8	0.275
Total SLR					0.15	–	–	0.2	0.1

Economic consequences include loss of goods and crops, the impossibility of working, and delays in transport, while the primary social consequence is increased poverty.

Each measure was assigned 0.025 points for each category of consequence, and therefore received a score of 0.075 for three consequences, up to a maximum of 0.1 points if all four types of consequences were found to result in the absence of that measure (Table 13.2).

According to that ranking, results indicate that drainage and its maintenance, the resettlement of residents in flood-prone areas to safe sites, and raised latrines are the most important measures given the consequences that their absence engenders. Next (with 0.075 points) come disaster preparation, vaccination campaigns after the event, hygiene education, stone barriers, and sea walls. With 0.05 points, the adaptation measures considered to be of lesser importance included early warnings and information, the identification of residents exposed to flooding, electric generators and drinking water tanks for hospitals, and the restoration of mangroves. Measures scoring just 0.025 points were disaster drills and emergency plans in schools (Table 13.2).

The inclusion of certain types of construction work in structural measures, rather than others, was reviewed following the interviews, and interviewees' opinions on the utility of measures were truly enlightening. For example, the most surprising result was the importance that interviewees attributed to drainage canal maintenance (Fig. 13.5) and rubbish removal.

Another aspect that emerged from the interviews was disappointment with resettlement operations for the residents of flood-prone areas. Though resettlement areas are not at risk of flooding, their peripheral locations often lack amenities and leave resettled flood victims far from their community and relatives, a tie considered to be very important in Maputo. Such was the case with the resettlement of central neighborhoods of the inner Guachene *bairro* to the most remote parts of the Katembe district, beyond the Umbeluzi River.

The total weighting of adaptation measures found in the various districts produced the following results: Katembe and KaChamanculo 0.275, KaMaxakeni 0.425, and KaMavota 0.8. As such, the district on the south side of the Umbeluzi River is the least adapted while the inner district on the border with the municipality of Matola is the best adapted. No district is therefore completely deprived of adaptive capacity. However, the existing measures are insufficient and this makes the risk higher (Fig. 13.6).

13.7 Future Adaptation Measures

A focus group consisting of six people was organized at the Municipality of Maputo Directorate for Environment. Each of the Directorate members is familiar with the problem of flooding, not only in a professional capacity, but also as

Fig. 13.5 Maputo, KaMavota district, 3 Nov. 2011. Stormwater drainage along Dona Alicia Street with efficiency compromised by vegetation (Tiepolo 2011)

private citizens often residing in exposed areas. The purpose of the focus group was as follows:

1. Identify adaptation measures for future flooding;
2. Categorize measures as short-, medium- or long-term;
3. Define the length of these three terms in months/years;
4. Establish the priority of the measures identified.

In addition to the district and neighborhood interviews, the focus group examined each of the 32 adaptation measures listed in the photo album, approving or rejecting them according to their appropriateness to local conditions. Appropriateness was determined on the basis of five criteria identified by the focus group:

- No regrets: the measure must have beneficial effects, even in the absence of climate change;
- Requirements: no legislative changes are needed to implement the measure;
- Community acceptance: the measure must be accepted by the local community;
- Budget: the measure's cost must be feasible given the current municipal budget;
- Cost-benefit: benefits of the measure must outweigh the costs of implementation.

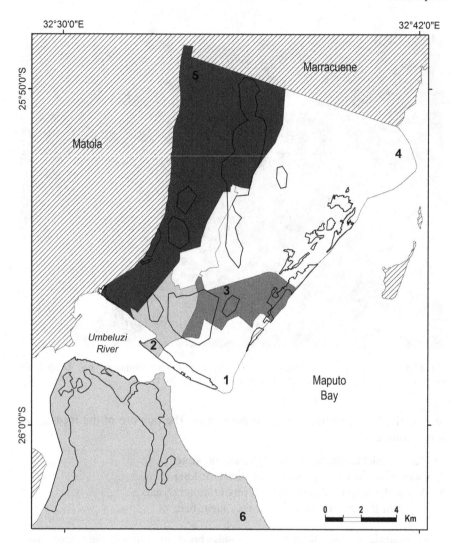

Fig. 13.6 Maputo 2011. Level of adaptation by district: most adapted is KaMavota (*5—dark grey*); moderately adapted is KaMaxakeni (*3—grey*); least adapted are Kachamanculo and Katembe (*2 and 6—light grey*). Exposed areas are enclosed by a continuous black border

Five of the 32 measures were simply considered inappropriate for Maputo without considering the 5 criteria. In particular, encouraging people to store food reserves was held to be irrelevant to the local urban environment. The emergency kit was deemed unbeneficial insofar as it might encourage residents to remain inactive during flooding. Drinking water tanks for hospitals were considered an unnecessary cost given that, to date, the water supply has never failed. Flood-proof platforms were considered to be a good measure for community services only.

Finally, resettlement was preferred over retrofitting to help homes withstand flooding (Table 13.3).

In addition, four measures required significant changes to legislation: protection of industrial water treatment plants, barriers to protect industries against flooding, an emergency plan for industries; and instructions on handling dangerous substances.

Five measures were considered beyond the reach of the municipal budget: slope stabilization, sea walls, and new stormwater drainage are all extremely costly. In addition, dune restoration has corollary economic implications given that builders use the dunes as a supply of inert materials for construction work. Finally, mangrove restoration is a costly and multifaceted endeavor because it requires educating local people regarding the consequences of cutting down the mangroves for firewood, and in some cases also necessitates resettlement.

Fifteen measures were considered short-term, four medium-term, and six long-term (Table 13.4).

Short-term measures are less costly to implement, thus all were retained without any prioritization. The 10 medium-long term measures were then pared down to six during the focus group discussion. Each member of the focus group was asked to score the six measures in order of priority from 1 (most important) to 6 (least important). The total points attributed by the focus group to each measure determined its priority (Table 13.5).

13.8 Implementing Adaptation Measures

The next step toward implementing the selected adaptation measures will be to involve the stakeholders from exposed areas of Maputo and any projects already in place to support community-based adaptation. These actors should be asked to re-orient their approach according to the 'no regrets' criterion (Care 2011).

Relocation to safer areas, improved housing, protection of productive assets, purchase of flood insurance, and flood preparation through training are all adaptation measures that are typically implemented at the household level (Moser and Satterthwaite 2008: 20).

In order to implement the measures contemplated here, physical planning tools should be suited to include climate change adaptation in urban environments. Indeed, such tools allow planners to focus on both peri-urban areas, where development can increase the hazards (such as flooding) that strike cities and downtown informal settlements, where most people live (Barros et al. 2012: 460). Physical planning tools affect land use, and thus the areas exposed to hazards and the physical aspects of vulnerability as well (United Nations 2004: 314). Though such tools have generally failed to include climate change adaptation in urban environments in the majority of developing countries (Barros et al. 2012), in some cases (e.g. Rufisque, Senegal) they have been adapted (Olhoff 2011: 29). In Niger, the national guide for drafting local development plans has been updated to

Table 13.3 Maputo 2011. The identification of priority measures for adapting to flood

Measure	No regrets	Legislative change	Community acceptance	Budget	Cost-benefit
Flooded area map at *bairro* level	•	•	•	•	•
Flood level markers	•	•	•	•	•
Early warning	•	•	•	•	•
Informing population about preparedness	•	•	•	•	•
Emergency preparation training	•	•	•	•	•
Emergency plans for schools	•	•	•	•	•
Disaster drills in schools	•	•	•	•	•
Health education	•	•	•	•	•
Evacuation map preparation	•	•	•	•	•
Incentives to leave exposed areas	•	•	•	•	•
Resettlement of exposed populations	•	–	–	–	–
Stormwater drainage maintenance	•	•	•	•	•
Identification of exposed population	•	•	•	•	•
Post-disaster vaccination campaigns	•	–	–	–	–
Power generators for hospitals	•	•	•	•	•
Protection of industrial water treatment plants	•	–	•	•	•
Anti-flood barriers for industries	•	–	•	•	•
Emergency plans for industries	•	–	•	•	•
Instructions on handling dangerous substances	•	–	•	•	•
Raised latrines above flood level	•	•	•	•	•
Concrete protection of wells against flooding	•	•	•	•	•
Dune restoration	•	•	•	Exp	•
Mangrove restoration	•	•	–	Exp	•
Slope stabilization	•	•	•	Exp	•
Dykes	•	•	•	•	•
Sea walls	•	•	Not all	Exp	•
New stormwater drainage construction	•	•	•	Exp	•

integrate climate change (République du Niger 2012). Nevertheless, *climatized* local development plans are presently few and mostly for rural areas (such as Gorouol, Loga, Roumbou, and Tanout), though the plan prepared for Niamey's district no. 1 shows that mainstreaming can benefit cities as well.

The process of mainstreaming climate change adaptation in national and sectoral policies and local planning tools is expected to guarantee that adaptation has a greater positive impact in the medium-long-term (UNDP 2011: 10).

Table 13.4 Maputo 2011. Measure time horizon: Short-term (0–6 months); Medium-term (6 months–1.5 year); Long-term (4 years)

Measure	Duration		
	Short-term	Medium-term	Long-term
Flooded area map at *bairro* level	•		
Flood level markers	•		
Early warning	•		
Informing population about preparedness	•		
Emergency preparation training	•		
Emergency plans for schools	•		
Disaster drills in schools	•		
Health education	•		
Evacuation map preparation	•		
Identification of populations exposed to flooding	•		
Post-disaster vaccination campaigns	•		
Protection of industrial water treatment plants	•		
Barriers to protect industries against flooding	•		
Emergency plans for industries	•		
Instructions on handling dangerous substances	•		
Dune restoration		•	
Mangrove restoration		•	
Raised latrines (above the flood level)		•	
Power generators for hospitals and health centers		•	
Slope stabilization			•
Sea walls and dykes			•
Stormwater drainage maintenance			•
Resettlement of populations exposed to flooding			•
New stormwater drainage construction			•
Economic incentives to leave exposed areas			•

Table 13.5 Maputo 2011. Medium-long term measure prioritization

Adaptation measure	Score	Ranking
Drainage maintenance	9	1st
Economic incentives to leave exposed areas	12	2nd
New storm water drainage	18	3rd
Sea wall	26	4th
Slope stabilization	27	5th
Resettlement	34	6th

13.9 Conclusions

The existing adaptation measures in Maputo are insufficient to significantly decrease hazard, vulnerability, and exposure to flooding. Not one of them has been properly designed to cope with extreme rainfall and high tides.

While the main rainwater drainage system can sustain the impact of even 200 mm of daily rainfall, the secondary network is small and obstructed by rubbish and vegetation in almost all areas.

The resettlement of residents from flood-prone areas to safe sites might seem like a simple and decisive measure, but it is implemented poorly and is disliked by the people benefiting from it who, over time, return to live in the areas they originally came from, where they can be close to their relatives and have access to basic amenities.

The situations in the four districts that encompass the areas at greatest risk of flooding differ enormously. While Katembe and KaChamanculo have almost no adaptation measures, KaMavota has three times as many. The risk equation for each area will therefore have a denominator that varies from 0.275 to 0.8, with significant effects on the final risk figure.

As regards future measures, those with the highest priority are mainly structural measures, particularly the maintenance of the existing rainwater drainage canal network and the construction of new sections. These findings are consistent with reports in the literature on other large African cities, such as Accra, Addis Ababa, Bangui, Dakar, Douala, Durban, Lagos, and Kampala (see Chap. 2).

The second priority measure involves the strengthening of banks, hillside slopes, and sea walls. Resettlement—even if it is implemented through economic incentives paid to those living in exposed areas—is considered to be of lesser utility.

As regards the methodology, ascertaining and weighing climate change adaptation measures requires discussion with local players, unlike other components of the risk equation. This is the only way to identify less visible adaptation measures and appreciate their relative importance. Secondly, adaptation assessment is more robust if the end goal is to take action. In fact, only when the issue of actually implementing measures is considered does the need arise to prioritize them, and thus to define the importance of single adaptation measures.

References

Allen A, Boano C, Johnson C (2010) Focus on adapting cities to climate change. DPU News 52, March: 2–4

Barros TF, Field OB et al. (2012) Managing the risk of extreme events and disasters to climate change adaptation. A special report of the intergovernmental panel of the groups and of the IPCC, Cambridge University Press, Cambridge

Braccio S, Tiepolo M (2013) Atlas des ressources locales: Kébémer, Louga (Sénégal), Niamey (Niger), http://cooptriangulaireins.net/

Care (2011) What is adaptation to climate change? Care International climate change brief, Oct

Dasylva S (2009) Inondations à Dakar et au Sahel. Gestion durable des eaux de pluie, Dakar

Davidson R (1997) An urban earthquake disaster risk index, The John A. Blume Earthquake Engineering Center, Department of Civil Engineering, Report No. 121, Stanford University, Stanford

FIFMTF (1992) Floodplain management in the United States: an assessment report, vol 2: Full report. L. R. Johnston Associates, Washington

Folorunsho R, Omojola A (2009) Lagos in World Bank framework for climate change risk assessment. Buenos Aires, Delhi

IMWI (2013) Strategic agenda for adaptation to urban water mediated impacts of climate change in Addis Ababa. Ethiopia

IPCC (2001) Climate change 2001: impacts, adaptation and vulnerability. Third assessment report. Cambridge University Press, Cambridge

Karley NK (2009) Flooding and physical planning in urban areas in West Africa: situational analysis of Accra, Ghana. Theor Empirical Res Urban Manage 4(13):25–41

Levina E, Tirpak D (2006) Adaptation to climate change: key terms. OECD, IEA, Paris

Lwasa S (2010) Adapting urban areas in Africa to climate change. The case of Kampala. Curr Opin Environ Sustain 2:166–173

Magadza CHD (2000) Climate change impacts and human settlements in Africa: prospects for adaptation. Environ Monit Assess 62:193–205

Moser C, Satterthwaite D (2008) Towards pro-poor adaptation to climate change in the urban centers of low-and middle-income countries, Global Urban Research Centre Working Paper 1

Nicholls R, Brown S, Hanson S, Hinkel J (2010) Economics of coastal zone adaptation to climate change. World Bank Discussion Papers 10

Olhoff A (2011) Opportunities for integrating climate change adaptation and disaster risk reduction in development planning and decision-making. Examples from Sub-Saharan Africa in Global assessment report on disaster risk reduction 2011, UNEP, ISDR

République centrafricaine (2009) Les inondations à Bangui République Centrafricaine. Evaluation de la situation actuelle et mesures pour réduire la fréquence et atténuer les impacts futurs. Rapport de l'évaluation conjointe des besoins, Sept

République du Niger, Min. du plan, de l'aménagement du territoire et du développement communautaire (2012) Document annexé au Guide national d'élaboration des PDC pour l'intégration de la dimension changements climatiques dans la planification communale

Tiepolo M (2012) Aménagement et gestion locale de l'environnement au Sahel in Tiepolo M (2012) Evaluer l'environnement au Sahel. Premières réflexions sur la gouvernance locale. L'Harmattan Italia, Paris-Turin

UNDP-UNEP (2011) Mainstreaming climate change adaptation into development planning. A guide for practitioners, Nairobi

United Nations (2004) Guidelines for reducing flood losses, http://www.unisdr.org

United Nations Habitat (2010) Climate change assessment for Maputo, Mozambique. A summary, UN-Habitat, Nairobi

Chapter 14
Linking Vulnerability and Change: A Study in Caia District, Mozambique

Elena Ianni

Abstract The vulnerability of a community is usually defined in relation to an outcome, such as hunger, following a natural catastrophe, like drought or flood. Present challenges demonstrate that emergencies are multi-faceted and that risks are not the result of physical events alone. Changes to ecological, economic or social relations and structures erode the adaptive capacity of communities, which become locked into an undesirable, vulnerable state. This chapter focuses on the ways in which a particular set of changes related to economic liberalization affect the vulnerability context in rural areas of the Caia District, in central Mozambique. The factors on which this study focuses include the modernization and resettlement policies of the district government; the market economy and its potential effects on agricultural incomes; and the land tenure investments of private firms and the increasing competition between farmers and commercial interests over access to land and natural resources. Drawing from household survey data and key informant data, we argue that the changes to economic and social relations that are occurring in the district may compromise families' capacities to cope with climate changes, and jeopardize their possibility of remaining on their existing development trajectory or making the transition to a better one.

Keywords Food security · Rural livelihoods · Subsistence economy · Land grabbing · Mozambique

E. Ianni (✉)
Department of Civil and Environmental Engineering, University of Trento,
Via Mesiano 77, 38123 Trento, Italy
e-mail: Elena.ianni@ing.unitn.it

S. Macchi and M. Tiepolo (eds.), *Climate Change Vulnerability
in Southern African Cities*, Springer Climate, DOI: 10.1007/978-3-319-00672-7_14,
© Springer International Publishing Switzerland 2014

14.1 Introduction

The answer to the question of whether human societies will be able to respond to the pace at which social and natural systems are currently changing is not trivial. Perhaps they can, if faced with either natural or social change (Anderies 2011). However, in the face of double exposure (i.e. the simultaneous impacts of global environmental change and globalization) (O'Brien and Leichenko 2000), households, villages, and regions may be stretched beyond their capacity to cope on their own. Both globalization and global environmental change are generating a highly connected system in which decisions at small scales are influenced by and will influence processes at the global scale in unpredictable and novel ways (Adger et al. 2009a, b). O'Brien and Leichenko (2000) and Leichenko and O'Brien (2002) suggest that there is a need to incorporate economic globalization into current analyses of climate vulnerability. In particular, the manifestations of globalization, including economic liberalization and increasing international trade, commercialization, and export orientation, are likely to shape future development patterns.

As explained by Silva et al. (2010), an examination of the dynamic interaction between environmental stress and economic change is particularly relevant for southern Africa, as it is characterized by high levels of both poverty and environmental variability. In addition, southern Africa is becoming increasingly integrated into the global economy and has been identified as one of the most vulnerable regions to climate change in the world. This chapter deals specifically with the vulnerability to climate change demonstrated by the rural Sena communities in the district of Caia in Mozambique, who exhibit all the aforementioned conditions characteristic of southern Africa, explored below.

Firstly, where poverty levels are severe, most people's food consumption is considered borderline according to international standards: it is below the recommended minimum daily calorie intake of 2,100 per person and falls below the recommended minimum daily protein intake of 52 g per person. In the inner part of Caia, there is insufficient access to or a general lack of basic infrastructure, such as roads and water wells. The mean per capita consumption of water in the villages is about 10 L/day. Smallholders' agricultural plots are rain fed and most families lose part of the harvest due to drought and locust invasions. Secondly, the district has increasingly been integrated into the global economy. In recent years, infrastructure development has attracted economic interests; the total population of the district increased by 35 % from 1997 to 2007. In order to improve and modernize rural living conditions (i.e. maximize safety and access to water), the government recently endorsed actions to relocate people from isolated rural areas to the neighborhoods of urban centers. Moreover, the district government, following a trend across Mozambique and in many other African countries, has been fostering large-scale land acquisitions and forest exploitation leases as a means to rapidly achieve an alleged rural development. These actions have been undertaken based on the assumption that they play an important role in catalyzing economic development, facilitating infrastructure expansion, and creating employment. The increasing competition with commercial interests over access to land and natural resources will force people towards the urban areas and will create large numbers of part-time, casual, and low-paid workers. Finally, environmental hazards, such as flooding, droughts, and cyclones commonly affect this vulnerable area. Repeated flooding of the Zambezi River has been documented in recent years, and evidence suggests that an increase in the variability and the intensity of rainfall events may be contributing to the associated flooding (Mason and Joubert 1997; Hulme et al. 2001; Thornton et al. 2006). Basically, climate stress will exacerbate the adverse ecological and socio-economic changes caused by large dams in the Zambezi catchment that have already substantially altered the magnitude, timing, duration, and frequency of flooding events in the delta (Beilfuss and Brown 2010). Projected climate changes—an increase in the mean annual temperature by 1.8–3.1 °C, a 2–9 % decrease in rainfall, and a 2–3 % increase in solar radiation by 2075—will lead to a drying in the Southern African region. Two outcomes are associated with these changes: a reduction in crop production, such as maize yields; and a decline in nitrogen content in plants, reducing their nutritional value.

The Sena communities that live in the Mozambican drylands are resource-dependent and have developed practices and institutions that are well adapted to local environmental variation. As a result, over the centuries families have developed a complex and generalized life system that has allowed them to survive in critical times and adapt to climate variability. In this chapter, I argue that the changes to economic and social relations that are occurring in the district may compromise families' capacities to cope with climate changes, and jeopardize their possibility of remaining on their existing development trajectory or making the transition to a better one. Drawing from household survey data and key informant data, I examine the effects and influences of simultaneous processes of change on farmers in the district of Caia. On the basis of the results, some policy actions in the context of evolving rural–urban interfaces and in the developing context of land deals are discussed.

14.2 Theoretical Framework

From a reductionist approach, people are considered vulnerable due to their presence in hazardous locations. The vulnerability of a household or community is usually defined in relation to an outcome, such as hunger, following a natural catastrophe, like drought or flood. Current challenges demonstrate that emergencies are multi-faceted (Eakin et al. 2009), and there is an emerging awareness that vulnerability and local responses to shocks and change can only be understood by examining the interactions between multiple stressors (Gallopin 2006; Vogel et al. 2007; Quinn et al. 2011). Vulnerability studies are finding increasing popularity in policy and practitioner communities engaged in various research fields, including risk assessment and natural hazards (Alcántara-Ayala 2002), food security (Grillo 2009), and environmental change (Adger et al. 2009a, b). O'Brien et al. (2007) maintain that there are two main streams of literature relating vulnerability to climate change. Authors define these two interpretations as *contextual* and *outcome* vulnerabilities. The latter considers vulnerability as the negative outcome of climate change for any particular exposure unit—an outcome that can be quantified and measured, as well as reduced through technical and sectoral adaptation measures, including by reducing greenhouse gas emissions. The former considers vulnerability as the present inability of social and ecological systems to cope with external pressures or changes (in this case, changing climate conditions) that are generated by multiple factors and processes. In this view, reducing vulnerability involves altering the context in which climate change occurs, so that individuals and groups can better respond to changing conditions. Vulnerability in drylands cannot be attributed to climatic factors alone but to a multiplicity of contextual conditions. This may include institutional and political changes, such as restricted access to land, water, services or mobility that often take place outside the realm of a rural community. Crop failures and livestock deaths are causing economic losses, raising food prices, and undermining food security with an ever greater frequency, especially in parts of Sub-Saharan Africa. Context vulnerability literature emphasizes that to effectively reduce vulnerability, factors that influence a generic capacity to respond to a variety of changes have to be identified (O'Brien et al. 2007). Identifying vulnerability to current variability and extremes is a good starting point for understanding vulnerability to any future changes in climate because some of the social factors that prevent certain households from responding today are likely to similarly affect households in the future.

14.3 The District of Caia

The district of Caia, Province of Sofala, lies in the lower tract of the Zambezi River basin. The population of the district is concentrated along the main communication corridor (i.e. the Beira-Sena national road along which the towns of Caia and Sena are located). Thanks to infrastructural improvements, along with the introduction of electricity, Sena and Caia have progressively assumed the shape and functions of rural towns (Diamantini et al. 2011). In the town, urban living occurs alongside rural ways of life, with subsistence agriculture and traditional institutions coexisting. These features reflect a spatial organization and housing typology much closer to that of a village than of a city. The interior part of the district still maintains a distinct rural character, and basic infrastructure—roads, schools, and water wells—are either lacking, or do not provide adequate services.

14.3.1 Sena Communities

The Sena people that live in the district are one of the Bantu speaking populations of Southern Africa. According to the classification of Sansom (1974), as western Bantu people the Sena of Caia concentrate their dwellings and disperse their economic activities over a wide zone of exploitation. This is based on ecological conditions, the main one being the nature of the terrain, with inland plateaus characterized by large expanses of relatively uniform country where a variety of resources are not frequently contained in small confines. Another determinative ecological condition is the general problem of finding a constant water supply source. The main activity for families is agriculture and, as in most African cultures, women are responsible for the agricultural work. Each family works on one or more agricultural plots, *machambas* (generally 3.5 ha per family), usually fragmented in different locations. One is set close to the house, between 5 and 30 minutes walking distance, and the others are located approximately 1 hour away, with some up to 2 hours or more from the family's house, in the lowlands (*baixa*). Fields in the lowlands are able to withstand periods of drought because they collect all the water available in the watershed. In the rainy season, the lower machamba is completely flooded, but the higher machamba keeps producing. In the lowlands, families are able to grow cash crops, including vegetables, sugar cane, and rice. The baixa produces about twice as much food as the rain-fed machamba. The main causes of harvest loss are drought and locust invasions. Basic widespread cultivated crops are sorghum, millet, and maize. Taking into consideration the cereal quantities available to families and assuming a consistent daily necessity of 0.5 kg of cereal per person, 50 % of families are not guaranteed adequate food.

In addition to basic crops, most of the families cultivate so-called high value crops (i.e. not directly used by the family) that are primarily intended for sale. Sesame is grown by the great majority of the families and is sold at 20 metical/kg (1 metical = 0.0363 US$). Cotton and peanuts are also widely grown in the district and are sold at 15 metical/kg and 6 metical/kg, respectively. Most of the time, agricultural production does not satisfy families' needs; therefore, families undertake temporary work (*buscados*) to survive during harsh periods. Families have learned how to cope with risk; their lives are built upon a complicated and delicate system, as well as different sources of income that span a wide territory and are highly influenced by chance. An expanded description of the communities in the district can be found in Ianni (2012) and Nicchia (2011).

14.3.2 Governance Issues

In 2006, the regional government drafted a strategic development plan for the district. Limited agricultural production, insufficient drinking water sources, periodic floods, and inadequate social infrastructure were identified as the main

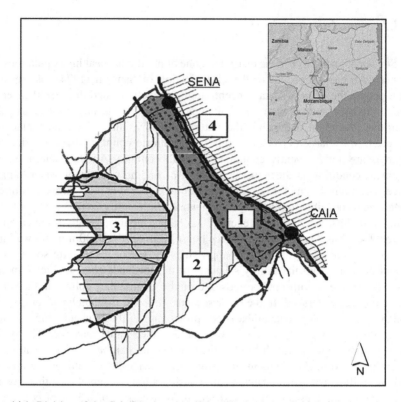

Fig. 14.1 Division of the Caia District according to the land use plan

constraints. On the other hand, the presence of abundant water sources (Zambezi River and Zangue River), high biodiversity in the forests, and soil fertility were described as the main strengths. On the basis of geographical and socio-economic criteria, the plan divided the Caia district into four areas (Fig. 14.1).

The government identified sub-region 1 as economically developed, sub-region 2 as a cluster of problematic communities, and sub-region 3 as a cluster of poverty. In the aftermath of the 2007 flooding of the Zambezi River, the government started a program to relocate the population from sub-region 4 to areas of lower risk (sub-region 1) in villages named *reasentamentos*. The rationale and expected outcome was that the Zambezi River fertile floodplain should become less settled. Also, the government recently endorsed actions to relocate people from isolated rural areas in sub-regions 2 and 3 to sub-region 1. These actions have been described by the government as a policy to maximize the population's access to water, schools, and markets. On the other hand, the idea of moving people from the forested interior sparked accusations of assisting large land investors to permeate these marginal and "wide" areas. The ambiguity of this policy derives from the fact that the land of the rural hamlets dispersed throughout the inner part of the district is not "underutilized", "uncultivated" or "available", as claimed by the government.

Meanwhile, contacts have been established between government officers in charge of agricultural matters and investors, but have not been made available to the public or non-governmental organizations.

Since land property in Mozambique belongs exclusively to the State, land may not be sold or alienated, but can be leased for the purpose of use and fruition (*direito de uso e aproveitamento*, DUAT). The Land Law, established in 1997, provides that DUATs are granted on a 50-year state leasehold, renewable for an additional 50 years. The law recognizes the *de facto* occupation of land by those occupying it according to customary norms and practices. Traditional land management systems continue to be the principal means by which the rural poor obtain access to land, as customary tenure presently accounts for 90 % of land tenure rights. Two suppositions in particular underpin the government's argument that land and forest deals are development catalyzers:

- Investors acquire the use of marginal or unused lands and increase their productivity and utility; and
- The award of concessionary rights to investors applying for land, according to the National Land Law, is subject to a mandatory consultation and negotiation process with the local community, and this is supposed to facilitate a win–win solution for local groups and investors.

14.4 Methods

Empirical analysis was undertaken regarding household vulnerability in the district of Caia to identify the key features that characterize the way in which climate stressors and economic changes interact with the local vulnerability context.

A vulnerability study that focused on spatial and temporal details at the individual or household level was gathered through a questionnaire conducted with families in the rural areas of the district. Questions were formally organized according to the following specific themes:

1. Food production (information on family organization, family agricultural yields, and non-agricultural activities that complement family livelihoods);
2. Food consumption (information on daily food intake and access to natural resources);
3. Food access (information on the type and amount of exchanges with the towns to acquire goods and commercialize production and living costs);
4. Adaptive capacity of families, i.e. response to uncontrolled events (information on the occurrence of agricultural calamities and strategies put in place by families to cope with crises); and
5. Farmers' relationships with the land (information on spiritual values and land attachment).

The qualitative sample did not intend to encompass all the families that inhabit the rural areas of the district; rather, a random sampling, influenced in part by accessibility, was adopted. The male or female heads of 72 households (with a mean of 8 family components) were interviewed. The age of the interviewees ranged between 18 and 84 years, with a median of 38. Thirteen individuals (10 of them women) were not able to specify their age. All interviews were conducted in the Chisena (indigenous) language. If the interviewee was able to speak Portuguese, open questions were discussed without translation.

Land access mechanisms were explored with the traditional authorities through in-depth interviews devoted to understanding the Land Law in practice. Questions posed to the traditional authorities comprised three themes. First, the customary mechanisms of land access for families were investigated to understand how land is requested and assigned. Then, authorities' knowledge of the Land Law and negotiation processes with investors, if any, were explored. Finally, the flow of information, within and outside the community was described to determine who makes decisions and how they are communicated to the community.

All interviewed authorities were male. In total, 26 interviews were conducted. No authority was able to speak Portuguese. Most of the authorities did not have a formal education, with the majority having completed schooling only up to the third grade. The mean age of authority figures was 57, with 8 under 40 years of age and 6 over 70 years old. The sample of authorities was representative, since it covered the totality of district *regulados* (administrative areas).

14.5 The Influence of Current Economic and Political Changes on the Vulnerability Context

The population of the district of Caia is facing multiple stressors. This investigation suggests that ongoing economic and political changes may impact the vulnerability context of households in terms of people's capacity to respond to current and future changes. We have analyzed the empirical data from the villages to assess some of the factors critical for response capacity. The factors on which we focus in this study include:

- The modernization and resettlement policies of the district government;
- The market economy and its potential effects on agricultural incomes, access to agricultural support, employment opportunities, state agricultural support systems, as well as household labor and capital availability; and
- The land tenure investments of private firms and the increasing competition between farmers and commercial interests over access to land and natural resources.

14.5.1 The Modernization Wave and Economic Growth

Many of the processes that generate vulnerability to climate change are closely associated with poverty. However, it has been demonstrated that poverty reduction measures may in some cases increase vulnerability and pose new threats to well being (Eriksen and O'Brien 2007). The transition from poverty to a less vulnerable state often fails when the social, economic, and political conditions that create uncertainty and insecurity are not removed. For example, migration and mobility are becoming increasingly important to generating income in response to climate change, and policies that tie people to small-scale agriculture and impose other constraints on mobility may, in fact, generate or exacerbate poverty.

In the district of Caia, relocation and urbanization actions are described by the government as a policy to modernize the district (i.e. maximizing public safety and access to water, schools, and markets). This policy followed a governmental analysis (Governo do Distrito de Caia 2006) that claimed the dispersion of settlements in the "wide" territory is the main cause of rural poverty and the primary obstacle to development, since it impedes capillary access to water and social facilities. The idea that economic growth is the essential tool for poverty reduction has been increasingly reconsidered, since it ignores non-income aspects of poverty and the processes of exclusion and marginalization that generate it. The current policy in the district is not addressing the real causes of poverty. For instance, the dispersed setting of the hamlets is not the cause of poverty; on the contrary, this pattern facilitates families' livelihoods and allows them to survive. The Sena people have built a successful risk management system. Diversifying access to fertile land is the strategy that the Sena people have always used, and which they believe to be the most effective action against climate change. The land considered by the government as the *remaining suitable land* for investments is not vacant, but actually already under use and extremely necessary to the livelihood strategies of rural people.

This research draws attention to ongoing resource access patterns and migration flows within the district. Consistent with the government analysis, the present study shows that, in the inner part of the district, there is insufficient access to or a lack of water wells. Of all the communities visited, only 25 % were found to have access to a functioning and easily accessible well. The remaining 75 % live with poor water quality in wells or difficult to no access to them. This is particularly true in the interior villages, where the problem of water scarcity is particularly severe: women need to walk more than 3 hours to reach a water source. As a consequence of new people relocating from the inner areas towards the corridor in search of water, access to resources (water, land, and fuel) in town neighborhoods is becoming difficult. In the villages close to the towns, wells do not have sufficient yield capacity to satisfy population needs and queuing time at the water wells is, on average, 1 hour. All the families living along the corridor walk 3–4 hours to a harvestable forest since closer forests have been overexploited, while 80 % of the families living in the interior collect wood in the machamba or in the vicinity of the house. Finally, along the corridor, fertile machambas are becoming scarce and people are returning to the mato in search of new machambas to cultivate.

I strongly assert that the proposed solution of reallocating people in the corridor to assure them a better life will fail completely if it is not coupled with projects to secure land for restoring forests and protecting water resources. Otherwise, urbanization policies may help in diminishing poverty levels, but will not reduce vulnerability.

14.5.2 The Shift to Commercial Farming

Transition to commercial farming for smallholder farmers requires at least two conditions: changing farming techniques and access to urban markets. However, farming techniques in the district that are well adapted to managing local environmental variability, such as seeding many small plots, are not suited to the economies of scale needed for commercial agriculture. In addition, although commerce in the entire inner district revolves around the city of Sena, the road network to reach the interior is in very poor condition and roads become impassable in the rainy season.

The government assumes that relocation actions will facilitate farmers' access to markets.

However, a critical finding in many studies is that, while people are dependent on market interaction, particularly during droughts, improving physical access to markets and increasing trade alone does not necessarily reduce vulnerability in the long term. This is because market access is not solely a function of geographical location in relation to regional and national markets or physical isolation. Equally important are the social and economic factors, which farmers in the Caia district lack, such as capital, support from agricultural authorities, and knowledge of market rules. All of these components place people in a poor negotiating position in market relations.

Government policies greatly benefitting commercial and private farmers at the expense of smallholders were established during the 1990s. In particular, commercial farms were granted access to bank credit, agricultural inputs, and extension services. Some of the large joint venture companies, formed when state farms were sold off after 1987, also contracted small-scale farmers, providing them with seeds and insecticides in exchange for the right to purchase, for example, cotton. The present relationship between producers and traders, in particular in the cotton sector, has not changed significantly, even since colonial times. As in the 1930s, during the monopolistic regime of the *Companhia Algodeira Nacional* (Isaacman et al. 1980), the factory imposes prices outside market rules. Under this system, contract smallholders with too little land may not even qualify for free inputs. In addition, smallholders who do not grow cotton or other crops for joint ventures often have no access to inputs at all.

In the wider district, very few families had received a visit from an agricultural extension agent at the time of data collection. In effect, the technological expertise provided through government programs has primarily benefitted wealthy farmers who can afford to invest in new technologies, while fewer technical options are available for rain-fed farming, or the plethora of other livelihood activities in which people engage (Eriksen and Silva 2009). Technological measures may actually be counterproductive and exacerbate the inequality and marginalization that currently contribute to the vulnerability of rural households, instead of effectively reducing it (Adger 2000).

Prices received for crops vary widely as a result of market forces. Since the price of cotton has drastically dropped in recent years and is currently around 5–6 metical/kg, many farmers abandoned the cultivation of cotton and shifted to growing sesame, which is sold at 20 metical/kg. The cost of agricultural products is directly proportional to their energy intake from the soil and the market considers sesame to have high nutrient uptake. This means that sesame cultivation is highly rentable because, theoretically, these returns can only last a few years due to the rapid depletion of soil fertility. The situation deserves particular attention from agricultural authorities. Sorghum, the crop that 89 % of the families cultivate, is taking a lot of energy from the soil and Caia farmers may soon face a long-term crisis caused by a decline in soil fertility due to thoughtless short-term decisions. On the other hand, price variability has not only increased the economic vulnerability of the people but also adversely affected their psychological disposition, as they find themselves embroiled in a situation over which they have no control.

14.5.3 Investments and Land Deals

The new Land Law of 1997 was designed to facilitate private enterprise as well as protect community rights; however, it may lead to adjustments in livelihood activities and affect people's capacity to respond to climatic fluctuations and change. It has become easier for investors and private enterprises to obtain land rights. While land cannot be sold or mortgaged, individuals and communities have the right to occupy their land, obtain a title document, and then rent it out on a 50-year lease. Peaceful conditions, as well as infrastructure upgrades since the end of the war, have resulted in an increase of foreign investments in Mozambique. Many studies argue that large-scale acquisitions jeopardize small-scale farmers' food security (Food and Agriculture Organization 2009) and access to the resources on which they depend, such as land, water, wood, and grazing (Haralambous et al. 2009; Kugelman and Levenstein 2009; Smaller and Mann 2009). This study in the Caia district is in complete accordance with these observations. The implementation of land deals will have a negative impact on the district.

This study found that local communities are totally unprepared to negotiate with emerging land investors, since awareness of their rights under the Land Law is very limited. There is a large and disturbing distance between state laws and practice

(Taylor and Bending 2009); shortcomings in the design and implementation of the community consultation process have been reported in the literature (Durang and Tanner 2004; Johnstone et al. 2004; Norfolk and Tanner 2007; Chilundo et al. 2005) and our results concur with these findings. In the district of Caia, communities that have been *consulted* and have reached an agreement often have no understanding that they are giving up this land permanently, and they have no understanding of the value of what they are giving away. Consultations were limited to one meeting and only community leaders (traditional chiefs, local party leaders) and a limited number of *community representatives* attended. It was rarely clear whether and how these individuals represented the community. The boundaries of the concession request were indicated on maps or in conversation but never verified (walked) during the consultation. Authorities completely ignored the rights the Land Law grants to local communities when negotiating with timber companies. Provisions of consultation records concerning benefit-sharing, time-bound targets or measurable indicators of progress could not be found, nor do proposed or effective sanctions exist in cases of non-compliance of the concession. All registered negotiations ended with short-term returns, often for the authorities and the participants of the meeting. In practice, several agreements between communities and investors involved one-shot decisions with a few people from the community accepting very small payments (a grain mill, a local shop) compared to the value of the forest concessions acquired by the investor.

The Land Law of 1997 established a mandatory community consultation process to pave the way for investors' access to land while protecting local land rights and improving living conditions. Although the explicit purpose of law is to protect and serve local communities, this does not seem to be the path to achieve that.

14.6 Conclusions

Climate change may amplify the cumulative effects of what communities are already experiencing, and has the potential to cause a collapse. To diminish vulnerability is most appropriately thought of as a process of social learning, using human capacities and knowledge to reduce vulnerability and risk in the face of the unknown and unexpected. There remains a contrast between individual and household strategies to cope and adapt, often relying on multiple sources of security, and the more sectoral and technical interventions promoted by national and international institutions. Policies and interventions should focus on the areas of overlap, such as adaptation measures that reduce both poverty and vulnerability to climate change. Unless the negotiating position of smallholders in market relations is improved concurrently with access to required technical expertise, inputs, and enhancement of land and natural resources for agricultural and off-farm activities, increased market interaction and commercialization can actually lock people into unreliable, short-term survival type strategies at the expense of long-term livelihood security and response capacity.

References

Adger WN (2000) Social and ecological resilience: are they related? Prog Hum Geogr 24:347–364

Adger WN, Dessai S, Goulden M, Hulme et al (2009a) Are there social limits to adaptation to climate change? Clim Change 93:335–354

Adger W, Eakin H, Winkels A (2009b) Nested and teleconnected vulnerabilities to environmental change. Front Ecol Environ 7(3):150–157

Alcántara-Ayala I (2002) Geomorphology, natural hazards, vulnerability and prevention of natural disasters in developing countries. Geomorphology 47:7–124

Anderies JM (2011) The fragility of robust social-ecological systems. Global Environ Change. doi:10.1016/j.gloenvcha.2011.07.004

Beilfuss R, Brown C (2010) Assessing environmental flow requirements and trade-offs for the lower Zambezi River and Delta, Mozambique. Int J River Basin Manage 8(2):127–138

Chilundo A, Cau B, Mubia M et al (2005) Land registration in Nampula and Zambezia provinces Mozambique. International Institute for Environment and Development, Maputo

Diamantini C, Geneletti D, Nicchia R (2011) Promoting urban cohesion through town planning. The case of Caia Mozambique. Int Dev Plann Rev 33(7):179–186

Durang T, Tanner C (2004) Access to land and other natural resources for local communities in Mozambique: current examples from Manica Province. Paper presented at Green Agri Net conference on 'Land registration in practice', Denmark, 1–2 April 2004

Eakin H, Winkels A, Sendzimir J (2009) Nested vulnerability: exploring crossscale linkages and vulnerability teleconnections in Mexican and Vietnamese coffee systems. Environ Sci Policy 12(4):398–412

Eriksen SH, O'Brein K (2007) Vulnerability, poverty and the need for sustainable adaptation measures. Clim Policy 7(4):337–352

Eriksen S, Silva JA (2009) The vulnerability context of a savanna area in Mozambique: household drought coping strategies and responses to economic change. Environ Sci Policy 12:33–52

Food and Agriculture Organization (2009) The state of food and agriculture. Biofuels: prospects, risks and opportunities. Food and Agriculture Organization, Rome

Gallopin GC (2006) Linkages between vulnerability, resilience, and adaptive capacity. Glob Environ Change 16:293–303

Governo do Distrito de Caia (2006) PEDD—Plano Estratégico de Desenvolvimento, Caia

Grillo J (2009) Application of the livelihood zone maps and profiles for food security analysis and early warning guidance for famine early warning systems network, USAID

Haralambous S, Liversage H, Romano M (2009) The growing demand for land. Risks and opportunities for smallholder farmers. International Fund for Agricultural Development, Rome

Hulme M, Doherty R, Ngara T et al (2001) African climate change: 1900–2100. Clim Res 17:145–168

Ianni E (2012) Land acquisitions and rural poverty: unveiling ambiguities in the district of Caia (Mozambique). Environ Nat Resour Res 2(3):52–61

Isaacman A, Stephen M, Adam Y et al (1980) "Cotton is the mother of poverty": peasant resistance to forced cotton production in Mozambique, 1938–1961. The Int J Afr Hist Stud 13(4):581–615

Johnstone R, Baoventura C, Norfolk S (2004) Forestry legislation in Mozambique: compliance and the impact on forest communities. Terra Firma Lda, Maputo

Kugelman M, Levenstein SL (2009) Land grab? The race for the world's farmland. Woodrow Wilson International Center for Scholars, Washington

Leichenko RM, O'Brien KL (2002) The dynamics of vulnerability to rural change. Mitig Adapt Strat Glob Change 7:1–18

Mason SJ, Joubert AM (1997) Simulated changes in extreme rainfall over Southern Africa. Int J Climatol 17:291–301

Nicchia R (2011) Planning African rural towns: the case of Caia and Sena. Lambert Academic Publishing, Mozambique

Norfolk S, Tanner C (2007) Improving tenure security for the rural poor Mozambique country case study. Food and Agriculture Organization, Rome

O'Brien KL, Leichenko RM (2000) Double exposure: assessing the impacts of climate change within the context of economic globalization. Glob Environ Change 10:221–232

O'Brein K, Eriksen S, Nygaard LP et al (2007) Why different interpretations of vulnerability matter in climate change discourses. Clim Policy 7(1):73–88

Quinn CH, Ziervogel G, Taylor A, Takama T, Thomalla F (2011) Coping with multiple stresses in rural South Africa. Ecol Soc 16(3):2. http://dx.doi.org/10.5751/ES-04216-160302

Sansom B (1974) Traditional economic systems. In: Hammond-Tooke WD (ed) The Bantu-speaking peoples if Southern Africa. Routledge and Kegan Paul, London

Smaller C, Mann H (2009) A thirst for distant lands foreign investment in agricultural land and water. International Institute for Sustainable Development, Winnipeg

Silva JA, Eriksen S, Ombe ZA (2010) Double exposure in Mozambique's Limpopo River basin. Geogr J 176(1):6–24

Taylor M, Bending T (2009) Increasing commercial pressure on land: building a coordinated response. International Land Coalition, Rome

Thornton PK, Jones PG, Owiyo T et al (2006) Mapping climate vulnerability and poverty in Africa. Report to the Department for International Development, UK

Vogel C, Moser SC, Kasperson RE, Dabelko GD (2007) Linking vulnerability, adaptation, and resilience science to practice: pathways, players, and partnerships. Glob Environ Change 17:349–364

Chapter 15
Conclusions

Silvia Macchi and Maurizio Tiepolo

Abstract The main lessons learned from the case studies presented in this book are reviewed in order to formulate recommendations for further research and action in adaptation to climate change. First, the present state of knowledge on the main components of vulnerability to climate change in large Sub-Saharan cities is assessed. To that end, the following set of components has been identified: hazard, sensitive area, exposure, adaptive capacity, adaptation measures, and risk. Next, a series of steps for improving available knowledge before or during planning adaptation is defined. Lastly, the issue of how to identify the most opportune adaptation options and measures is examined and a five-step process for planning adaptation is proposed as follows: 1. Identify possible measures; 2. Localize measures; 3. Plan measures; 4. Manage measures; and 5. Identify entry points. Challenges and related recommendations are outlined for each step of the process, with reference to the case studies of Dar es Salaam, Tanzania, and Maputo, Mozambique.

Keywords Urban Africa · Climate change hazards · Adaptation planning · Risk assessment · Vulnerability assessment

S. Macchi (✉)
Department of Civil, Building and Environmental Engineering, Sapienza University of Rome, Via Eudossiana 18, 00184 Rome, Italy
e-mail: silvia.macchi@uniroma1.it

M. Tiepolo
Interuniversity Department of Regional and Urban Studies and Planning, Politecnico di Torino, Viale Mattioli 39, 10125 Turin, Italy
e-mail: maurizio.tiepolo@polito.it

S. Macchi and M. Tiepolo (eds.), *Climate Change Vulnerability in Southern African Cities*, Springer Climate, DOI: 10.1007/978-3-319-00672-7_15, © Springer International Publishing Switzerland 2014

15.1 Current State of Knowledge and Perspectives on Improvement

The analysis developed in the two introductory chapters and in those dedicated to individual case studies has observed the state of knowledge on the main physical and human components that influence vulnerability to climate change in large Sub-Saharan cities: climatic and non-climatic drivers of hazard, exposure, sensitive areas, inhabitants' adaptive capacity, on-going adaptation measures, and risks. These components assume different denominations and are placed in relation to each other in different ways, depending on the study question and disciplinary context as well as the approach to interpreting vulnerability (what matters) and defining the consequent strategy for action (what should or could be done). The book moves in the direction of a shared reference framework. However, deeming that the time is not yet ripe for final methodological choices with respect to planning adaptation to climate change, the editors preferred to leave the author of each chapter free to adopt the terminology best adapted to his or her particular case and theoretical approach, asking that they simply outline their reference frameworks within the text. Moreover, this decision made possible the types of comparison and contamination that are necessary when addressing Sub-Saharan planning culture, which was born of colonization and is today subject to processes of continuous hybridization as a result of the interventions of donors and consultants from all over the world.

In this book, the question of vulnerability to climate change is addressed through two main methodological perspectives: that of adaptation to climate change in the case of progressive salinization of groundwater due to seawater intrusion in Dar es Salaam, and that of disaster risk management in the case of flooding caused by extreme rainfall events and sea level rise in Maputo. The IPCC Special Report on Managing the Risks of Extreme Events and Disasters to Advance Climate Change Adaptation (2012) recognizes that these two approaches have a shared focus on reducing exposure and vulnerability to natural hazards, including the potential adverse impacts of climate change. Nevertheless, the adaptation perspective defines vulnerability as a dependent variable that is determined by contextual factors such as exposure, sensitivity, and adaptive capacity, while the disaster risk reduction perspective defines vulnerability as an independent variable that is combined with exposure and adaptation in the determination of risk.[1] It bears mentioning that the most recent policy literature on adaptation to climate change reveals a tendency to consider these two approaches as closely interrelated,[2] although adaptation includes aspects that go beyond the mitigation of

---

[1] In fact, the disaster risk reduction definition of vulnerability appears rather similar to what the adaptation approach defines as sensitivity.

[2] For example, the 2013 EU Strategy on adaptation to climate change states "Adaptation action is closely related and should be implemented in synergy and full coordination with the disaster risk management policies that the EU and the Member States are developing".

possible adverse impacts of climate change to include the consideration of opportunities provided by a changing climate.

The main lessons learned that have been presented in this book are reviewed below and grouped according to a list of the components of vulnerability. Per the above comments, the list itself should be considered merely a tool used for the summary purposes of this conclusive chapter.

Hazard

This book has outlined the types of analysis that can allow us to understand whether climate change is under way and to what extent highly weather- and climate-dependent hazards can be attributed thereto, such as seawater intrusion caused by the combined effect of changing rainfall patterns and sea level rise or tidal surges. The most significant analyses are those that address extreme rainfall and the daily rainfall return period (see Chap. 9) as well as variations in minimum and maximum temperatures and tidal height and speed (see Chap. 3). Yet such studies are rarely carried out (see Chap. 2). The necessary information is often unavailable, as is frequently the case with rainfall intensity (mm/h), or has not been collected over a sufficiently long period of time, as with tidal gauge data (m/h), and/or is not adequately detailed, as is the case with variations in daily rainfall values over the course of a year. However, the case of Maputo (see Chaps. 9 and 10) demonstrates that even with the information presently available, one can conduct extremely useful analyses that allow for the identification of increased extreme rainfall frequency over the last decade. But the fact remains that analyses of the entire river basin within which the city lies is recommendable, as precipitation occurring far from the city may cause the river to swell and flood (see Chap. 2). In addition, hazards result from the combination of climatic and non-climatic factors such as urban sprawl, particularly in areas of rapid urbanization (see Chap. 5). It follows that analyses of climate and weather parameters must be accompanied by analyses of land cover and urban development patterns. Moreover, the relationship between hazards and climate conditions is not always as direct and immediate as in the case of flooding and heavy rainfall. The phenomenon of seawater intrusion into coastal aquifers, for example, represents a much more difficult hazard to assess since it is the result of a complex dynamic involving multiple variables and medium-long time periods (see Chap. 4).

Sensitive Areas

The possible assessment methods vary on a case-by-case basis. In the case of flooding due to heavy rainfall, the direct identification of hazard-prone areas is rapid, precise, and not very costly. In fact, post physical event direct surveys are already practiced by many local administrations in order to assess damages (see Chap. 2). However, this can still be improved. First and foremost, damage survey data should be entered into a GIS and placed in relation to the intensity of the precipitation that caused it (see Chap. 11). Direct survey, insertion of data into a specific GIS, and study of the relationships between variations in rainfall patterns—as well as a series of anthropic factors that modify water extraction and recharge rates—are also essential steps in identifying the areas sensitive to seawater intrusion (see Chap. 4). In both cases, the need for research into expedited survey methods based on remote sensing and ground monitoring networks is more urgent than ever.

Exposure

In the cities studied, the sensitive areas are vast (see Chaps. 3 and 11) and contain a large number of people (130,000 in Dar, 70,000 in Maputo). Rapid methods for estimating the resident population during inter-census periods (see Chap. 5) and evaluating the sensitivity of livelihood systems to environmental changes (see Chaps. 6 and 10) represent a priority for urban research. However, in the peri-urban fringe, many flood-prone areas contain few or no inhabitants. As such, according to the current definition of hazard, they would not be taken into consideration. Yet these areas are often intensively cultivated and constitute a source of sustenance and income for many citizens. It is therefore essential that exposure assessments be extended to include peri-urban agricultural areas.

Adaptive Capacity

In the case of Sub-Saharan cities, adaptive capacity is an interpretive key that allows for a positive view of a reality that is often described exclusively in terms of what it lacks compared with 'real' cities (see Chap. 1). Investigation of household characteristics can contribute to an understanding of how different modalities of resource access and environmental management, spatial location, facilities, and economic activities might promote or limit specific autonomous adaptation practices. The field study conducted in Dar es Salaam's peri-urban areas (see Chap. 6) reveals a broad spectrum of options (explicit and implicit adaptation practices) from which institutions can draw valuable lessons in terms of identifying potential entry points for effective institutional adaptation measures. Nevertheless, the proposed framework for assessing autonomous adaptive capacity requires further study in order to include analysis of the institutional dimensions of adaptation, thus providing a basis upon which to assess the effectiveness of adaptation policies and actions.

Adaptation Measures

The case of Maputo (see Chap. 14) demonstrates that assessing the climate proofing measures in place is very complex and time consuming even for visible measures, given the extent and fragmentation of sensitive areas, such as those prone to flooding. The large cities south of the Sahara rarely have georeferenced maps of the drainage networks or their level of maintenance. Censuses of houses are obsolete and refer to a territorial unit that is too vast to be of any use. There are no maps of slope stabilization works or water regulating structures like embankments and dams. Even more complex is the identification of measures taken to improve the population's autonomous adaptive capacity, as these rely on multiple sources of security and require interventions that go beyond and even conflict with the more sectoral and technical measures usually promoted by institutions (see Chap. 6). Learning from local communities is fundamental when defining opportunities and implementing useful adaptation pathways (see Chap. 8).

Meanwhile, an effort must be made on the part of local governments to mainstream climate change concerns in their daily work rather than focusing on self-standing measures (see Chap. 7).

Risk

Risk mapping is the first and most essential step in the spatial planning of adaptation. But, given the continuous transformation occurring in the large cities south of the Sahara, there is a need to shift from the use of one-time mapping to the concept of continuous monitoring. Mapping must be precise, i.e. conducted on a detailed scale, and at the same time simple to calculate. The use of GIS is thus indispensable, and the flood risk assessment carried out in Maputo is an example of good practice (see Chap. 12). Nevertheless, the coexistence of a plurality of risks in the same area, as is the case in Maputo where floods are due to both heavy rains and sea level rise, should be taken into account. As such, a multi-risk assessment should be developed, one that draws on inputs from several different disciplines.

Without knowledge it is difficult for local administrations to effectively address adaptation to climate change, even if they are sensitive to the problem and have access to the necessary resources. Available knowledge influences the choice of adaptation measure. A series of complementary and in some cases preparatory steps for planning adaptation has thus been identified:

- Improve data acquisition networks and monitoring systems;
- Georeference the data obtained and insert them into a multiscope GIS;
- Identify hazard trends under climate change;
- Reinforce local administrative capacity to produce and manage information;
- Create incentives for ground-up knowledge building.

15.2 Towards Adaptation Planning

The purpose of our investigations has also been to access as many elements as possible in order to identify, on a case-by-case basis, the most opportune adaptation options and measures. The nature of adaptation measures varies considerably according to the chosen approach: action-specific, mainstreaming, or a combination of the two (see Chap. 7). This choice depends in turn on a variety of interconnected factors: the type of phenomenon under consideration (e.g. flooding from extreme rainfall or progressive groundwater salinization due to incremental shifts in rainfall and sea level), the time horizon of the intervention (short or medium-long term), the scale of the intervention (from a single neighborhood to the entire urban region), the components of risk or vulnerability targeted by the intervention, the definition of adaptation used (which ranges from mere reduction of potential climate-related development risks to taking advantage of opportunities), and the budget and policy priorities of the local administrations involved. In this book we have outlined two case studies that differ considerably in terms of these factors, and as such the action-specific approach prevailed in the case of

Maputo while the mainstreaming approach was adopted in the case of Dar es Salaam. However, we contend that in both cases the transition from knowledge of phenomena to adaptation followed the same five main steps, outlined below.

Step 1: Identify Possible Measures

Whatever approach is adopted, a careful analysis of the physical, social, and institutional characteristics of the city in question is fundamental to defining a series of criteria with which to evaluate the appropriateness of each measure.

In the case of Dar es Salaam, the measures studied were aimed at transforming existing local plans for urban development and environmental management in order to increase their capacity to support the inhabitants of peri-urban areas in their autonomous adaptation to environmental changes (see Chap. 7). This is a typical mainstreaming measure, which entails changes to the procedures and organization of local institutions, the normative apparatus, and policy frameworks in order to transform decision-making processes such that the structural and non-structural interventions undertaken by administrations contribute to improving the adaptive capacity of the settled population. Several important lessons can be drawn from such on-going activities. First, the experience that local administration gained through working on the application of the Environmental Policy Integration (EPI) principle, particularly when developing Agenda 21 in a local context, has provided a fundamental basis of competence and capacity in mainstreaming adaptation to climate change. In fact, in the case of Dar es Salaam such experience has been so important that, today, capacity building in adaptation mainstreaming is often a question of recognizing local administrations' existing contributions to adaptation rather than introducing a new approach. The EPI work undertaken thus far has also prompted a series of unresolved tensions that also recur with respect to adaptation. In particular, there is a risk that the introduction of new procedures like those typical of the EPI model (Environmental Impact Assessment and Strategic Environmental Assessment) may translate into a reinforcement of state control over local governments. That risk is particularly pronounced in cases of urban adaptation where there is already an open conflict between national interests in the city as a motor of economic growth and the interests of the majority of the urban population, which sees the city as a resource for achieving their own life plans. The two parties in question have extremely different concepts of what urban assets and processes should be protected and enhanced through adaptation. Guaranteeing that these two conceptions compete on equal footing and reach an equitable compromise in terms of what to do entails that planners assume the important responsibility of opening decision-making processes up to the participation of inhabitants.

In the case of Maputo, assessments addressed two typologies of intervention that could be implemented by local institutions: structural and non-structural (see Chap. 13). As regards the former, a clear preference emerged for small-scale work that could be managed by local communities. The most controversial intervention was evacuation of the population from the most exposed zones. The practicability of evacuation depends on the capacity of local governments to offer valid resettlement alternatives for families and activities through proven instruments such as the use of vacant land and land sharing. Such instruments require a georeferenced information system (GIS) regarding land ownership in order to assess whether the existing unoccupied urban spaces are large enough to accommodate the transfer. In addition, it is essential that measures be taken to prevent future occupation of risk areas, which often occurs when a risk area has neither a use nor a user. Floodplains can serve many purposes and be used for a variety of purposes as long as they are not transformed by physical structures that would impede their hydraulic functionality. Among non-structural interventions, the use of early warning systems appears to be widely practicable. Nevertheless, its efficacy depends on the correct identification of potentially disastrous events and the continuous updating of such knowledge so that the alarm is modified over time. The case of Maputo also demonstrates that structural and non-structural interventions must be integrated: for example, the development of new drainage canals would be ineffective if not accompanied by adequate waste management (see Chap. 13).

Step 2: Localize Measures
The localization of measures and the scale of intervention depend on the type of problem being considered, the chosen adaptation strategy (climate proofing versus adaptive capacity improvement, see Chap. 7) and the type of measure to be developed.

In the case of Dar es Salaam, the problem is the vulnerability of access to freshwater for households dependent on boreholes within the coastal plain, where the combined effects of climate change and urban expansion are expected to aggravate the salinization level of the shallow aquifer. The adaptation strategy aims at improving the autonomous adaptive capacity of households through adoption of typical mainstreaming measures in order to incorporate adaptation into local urban development and environmental management plans, as mentioned above. This is therefore a local problem in a clearly defined area—the Dar coastal plain—and interventions target the plans of local administrations that impact important variables of the problem (depending on the case, these may be the municipal development plans, metropolitan master plans, or basin management plans) as well as involving advocacy for changes in norms or policies among higher level institutions.

In the case of Maputo, the problem is flood risk in low-lying areas due to extreme rainfall events, which are expected to increase as a result of climate change. The adaptation strategy is to climate proof the city through a variety of structural and non-structural action-specific measures. Such measures do not concern exclusively the area most exposed to risk. Rather, in the view of prevention, the basin uphill from the city is also considered in order to reduce runoff and erosion before they can generate impacts on the city. This involves interventions in territories administered outside the city, and actions differ from those implemented in the city. This may involve the transition to more energy efficient cooking stoves, the use of alternative fuels such as pellets to prevent deforestation, land and water conservation works to promote infiltration and reduce runoff (trenches, embankments, stone walls, etc.), community management of grazing land, or the cultivation of crops that do not necessitate clear cutting. None of this has anything to do with the lack of urban drainage canals, which requires action by other institutional actors, but it does relate to a decrease in pressure on existing river levees and drainage systems. In order to achieve these adaptations, some intermediate administrative entity, such as the region, must promote dialogue between the local administrations in question in order to sensitize them to the problem and then modify their respective physical planning instruments.

Step 3: Plan Measures

Once the set of possible measures has been defined and the relevant public administrations have been identified, interventions must be planned and distributed over time according to a clear list of priorities and with a clear understanding of the costs and modalities of implementation. In other words, measures must be inserted and harmonized within a local adaptation action plan. While many Sub-Saharan countries have developed this type of instrument at the national level (National Adaptation Programme of Action—NAPA), the number of large cities that have a Local Adaptation Plan of Action (LAPA) remains low. Moreover, existing LAPAs differ considerably, oscillating between true plans for adaptation to climate change to climate-compatible or climate-proof development plans (see Chap. 2).

It bears mentioning that in both Dar es Salaam and Maputo, many of the measures identified above have already been included in local planning for some time as they are considered essential to achieving the MDGs (see basic infra-structure) and application of the EPI principle (see Strategic Environmental Assessment). Nevertheless, this is not enough to guarantee that they be imple-mented, even in the presence of external aid in the form of financial resources and capacity building. Today, many of the efforts to overcome the gap between planning and implementation are oriented towards improving the budgetary self-reliance of local governments in order to reduce their financial dependence on donors and increase their ownership of policy implementation. One of the most widely used strategies is the reinforcement of local fiscality, a very delicate argument since it draws attention to the above-mentioned conflict between national interests and those of inhabitants, in addition to opening the question of land tenure and titling and the taxation of land rents. In practice, it is a matter of improving knowledge of taxable income, simulating earnings, increasing the collection of taxes, and using tax earnings to finance climate change adaptation measures and planning instruments in a manner that guarantees cost effectiveness and efficiency in public spending. In many cases, the budgetary decision-making processes are also being reviewed, responsibility mechanisms have been introduced into spending decisions, and the role of local governments' political branches, partic-ularly that of the mayor, have been reduced in favour of the administrative branch. Much work has yet to be done with respect to improving efficiency and equity in spending, however, and there is also a high risk that a shift towards technocracy may begin to occur.

In such a context, planners can assume a pivotal role. While the importance of spatial planning is evident with respect to the location of taxable assets and activities (as in resettlement operations and/or the use of vacant land), one must not forget that planning is also a strategic tool for capturing at least part of the surplus value of the land generated by public works such as new infrastructure (paved roads equipped with drainage canals and streetlights) and urban facilities (schools, health care, shopping centers, markets, etc.). The capture of this surplus value (e.g. through taxation of vacant private land or leasing of vacant public land) can contribute to financing climate change adaptation work.

Step 4: Manage Measures

The adaptation measure management phase, even in the presence of a good action plan, remains a challenge.

How can the maintenance of improved infrastructure be guaranteed? Existing assets are managed poorly or not at all. Certain types of work can be entrusted to community management if there is an incentive supply. For others, however, it is difficult to imagine anything other than improved public management. As always, it is a question of means (personal, instrumental, and financial) and institutional capacity. The problem is not much different when it comes to ensuring the application of new procedures, the functioning of new organizations, the enforcement of new laws, or the adoption of new policy frameworks.

Added to this are the management requirements specific to adaptation to climate change. Notwithstanding the fact that in many cases the quantity and quality of available information could be substantially improved by acquiring new data or simply by analyzing existing data in a more effective manner, it is undeniable that uncertainty is inherent in predictions of global climate change, its local effects, and above all its impacts on natural ecosystems and human livelihoods (see Chap. 1). As a result, very few adaptation decisions can be considered irreversible, and local administrations need the capacity to continuously revise strategies and interventions on the basis of the evolving availability of information. While the absolute priority of no or low regret measures remains, the management phase should be considered an integral part of adaptation action, and decision-making processes should be endowed with the flexibility necessary to take into account the dynamic nature of the risks associated with climate change and the vulnerability of the contexts in question.

Step 5: Identify Entry Points

This is the most complex phase, both because there is very little pre-existing experience upon which to rely and also because experiences obtained elsewhere may provide very limited support in terms of the choice of so-called adaptation entry points. The place-specific variables that influence such choices are manifold: certainly the particular society-environment relationship that characterizes a city and determines sensitivity to a given effect of climate change plays a key role; the "planning culture" of a given society is equally relevant, as it constrains the choice of modality through which the planning process can occur as well as the type of approach that is effectively practicable; finally, recent emergency situations can sensitize public opinion with respect to a specific risk, or an existing urban policy could be useful for adaptation and thus financed with climate change funding.

Today, the most widespread strategy is to concentrate on activities and measures that are also justified under current climate conditions and able to provide short-term tangible benefits to well-defined target groups. This does not prevent one from assuming a broader perspective and undertaking actions aimed at reinforcing the general capacity of local administrations with respect to environmental management and urban development planning, as the cases addressed in this book have demonstrated. For example, contributing to the development of a multiscope information system could be a next step as it offers multiple advantages that the administrations of large Sub-Saharan cities appear to appreciate. In this context, bilateral cooperation with the direct involvement of non-state actors (universities, local administrations, NGOs, CBOs, etc.) emerges as a modality capable of offering good results, and should therefore be prioritized.